SPALTING 101

The Ultimate Guide to Coloring Wood with Fungi

DR. SERI C. ROBINSON

SCHIFFER PUBLISHING

4880 Lower Valley Road · Atglen, PA 19310

Other Schiffer Books by the Author:

Living with Wood: A Guide for Toymakers, Hobbyists, Crafters, and Parents, with photomicrographs by Dr. Sarath M. Vega Gutierrez, ISBN 978-0-7643-5935-4

Spalted Wood: The History, Science, and Art of a Unique Material, with Hans Michaelsen & Julia C. Robinson, ISBN 978-0-7643-5038-2

Other Schiffer Books on Related Subjects:

New Woodturning Techniques and Projects: Advanced Level, Helga Becker, Photography by Richard Becker, ISBN 978-0-7643-5018-4

Home Woodworker Series: Home Workshop Setup—the Complete Guide, Jim Harrold, ISBN 978-0-7643-4418-3

Copyright © 2020 by Seri C. Robinson

Library of Congress Control Number: 2020931235

Cover and interior designed by Ashley Millhouse
Front cover image: *Legacy* by Mark Lindquist and Seri Robinson, a collaborative work out of spalted elm
Type set in Capitolium

ISBN: 978-0-7643-6089-3
Printed in China

Published by Schiffer Publishing, Ltd.
4880 Lower Valley Road
Atglen, PA 19310
Phone: (610) 593-1777; Fax: (610) 593-2002
E-mail: Info@schifferbooks.com
Web: www.schifferbooks.com

For our complete selection of fine books on this and related subjects, please visit our website at www.schifferbooks.com. You may also write for a free catalog.

Schiffer Publishing's titles are available at special discounts for bulk purchases for sales promotions or premiums. Special editions, including personalized covers, corporate imprints, and excerpts, can be created in large quantities for special needs. For more information, contact the publisher.

We are always looking for people to write books on new and related subjects. If you have an idea for a book, please contact us at proposals@schifferbooks.com.

*To Bill Wiard, the guy from the backwoods of the
Upper Peninsula of Michigan who convinced me
that people would want to hear about my
spalted wood research.*

"NO, SEE, THE BEER IS FOR DRINKING AND *THEN* YOU URINATE ON THE WOOD. IT'S GOOD FOR THE SPALTING."

—an unnamed spalted wood fan

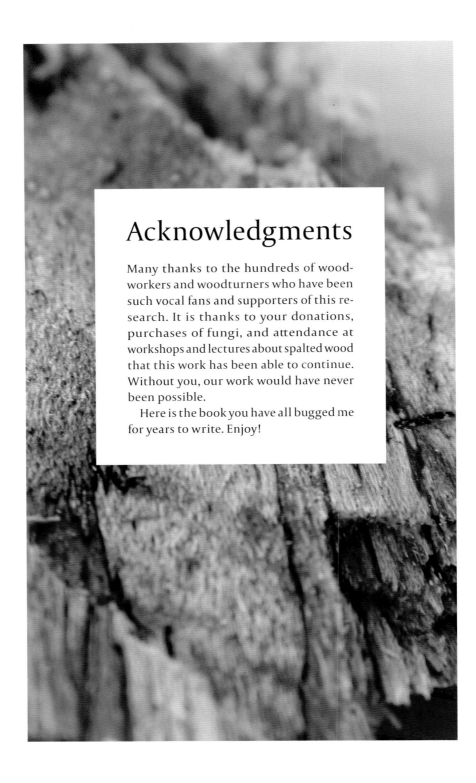

Acknowledgments

Many thanks to the hundreds of wood-workers and woodturners who have been such vocal fans and supporters of this research. It is thanks to your donations, purchases of fungi, and attendance at workshops and lectures about spalted wood that this work has been able to continue. Without you, our work would have never been possible.

Here is the book you have all bugged me for years to write. Enjoy!

Contents

Introduction   8

1. Basics of Wood Anatomy   10
2. Basics of Fungal Anatomy   28
3. Debunking Myths   38
4. Zone Lines   42
5. White Rots   55
6. Pigments   61
7. Combinations   84
8. Home Cultivation of Fungi   86

Appendix I.
Common Spalting Fungi and Where to Find/Buy Them   92
Appendix II.
Best Woods for Spalting   102
Appendix III.
Notes on Woodworking with Spalted Wood   104
Appendix IV.
Notes on Woodturning with Spalted Wood   106
Appendix V.
Most-Common Questions and Their Answers (FAQ)   110
Appendix VI.
Spalting Appearance by Wood Species   112

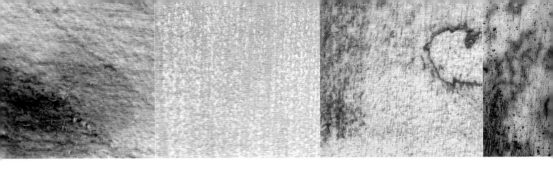

Introduction

Back in 2014, when I was just putting the finishing touches on *Spalted Wood: The History, Science, and Art of a Unique Material*, I thought, *this is it! This is everything everyone needs to know about spalting!*

Turns out I was wrong.

Spalting is so much more than its history or current use. Spalting is a complex intersection of science, art, history, and human development that could take a lifetime to truly unravel and properly explore. Since 2014, we (meaning myself, my students, and my collaborators) have found spalted work all over the world, both historical and modern. From ancient legends to modern sculpture, spalted wood continues to fascinate, but what I get asked more often than anything else is *Yeah, but how do you DO it?*

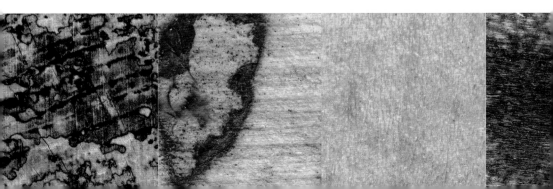

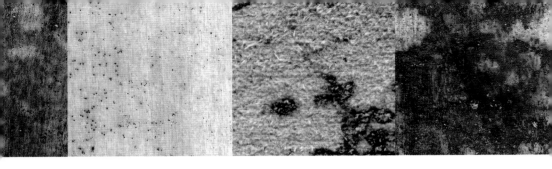

This book gently sets aside history and modern context in favor of straight-up biology—what spalted wood is in its most basic components, how it interacts with wood, and how it can be induced and controlled. All aspects of the biology are touched upon, from wood and fungal anatomy, to decay, to spalting recipes and safety tips for woodworking. An easy FAQ at the back offers an even quicker way to get answers to some of the most prevalent questions about spalting.

If you're new to spalting, welcome to the rabbit hole. Even those who think they know spalting will find this deep dive into spalting biology fascinating. While it's meant for a lay audience, those interested in the more subtle decay and mechanical issues associated with spalted wood will not be disappointed.

And for the DIYers, this is the guide you've long been promised.

Go forth, friends, and spalt!

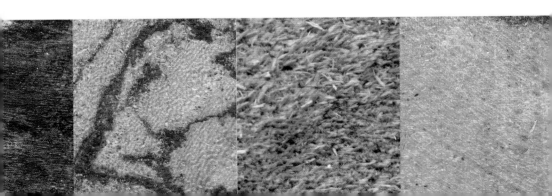

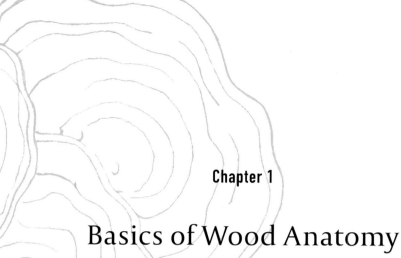

Chapter 1

Basics of Wood Anatomy

At its most basic level, wood is a collection of "straws" that run along the long axis of the tree (up and down), with a smaller number that run 90 degrees to that long axis ("in and out"). Inside the cells (you can keep thinking of them as straws if that helps) are various other components, such as wood extractives, that lend some additional strength and durability properties to wood. This chapter takes a broad look at the anatomy of temperate hardwoods and softwoods, as well as tropical trees, in terms of their cell types, arrangements, and extractive contents, as a way to better contextualize how fungi move through wood, and why some woods spalt better than others.

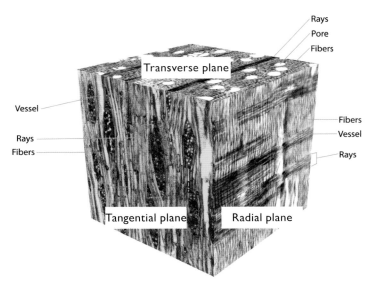

3-D micrograph of an Oregon maple (*Acer macrophyllum*) cube, showing the transverse, tangential, and radial planes. *Image courtesy Dr. Sarath Vega Gutierrez*

TEMPERATE HARDWOOD

A hardwood tree is a deciduous tree, so it drops its leaves in the fall. There are, of course, deciduous conifer trees, but we will ignore the outliers for the purposes of this book. "Hardwood" and "softwood" have nothing to do with the density of the wood and are entirely based on the classification system.

Temperate hardwoods—those growing in temperate regions that experience seasons, such as North America, the United Kingdom, Russia, etc.—show many similar features. They are composed primarily of **vessel elements** (also known as "pores"), which are used for conduction and run longitudinally; **fiber tracheids**, which provide support and also run longitudinally; **parenchyma**, which are used for storage; **rays**, which run at 90 degrees to the long axis (or *radially* from the pith); and occasionally **tyloses**, which are parenchyma cells that have expanded into a vessel element and cause blockage.

11

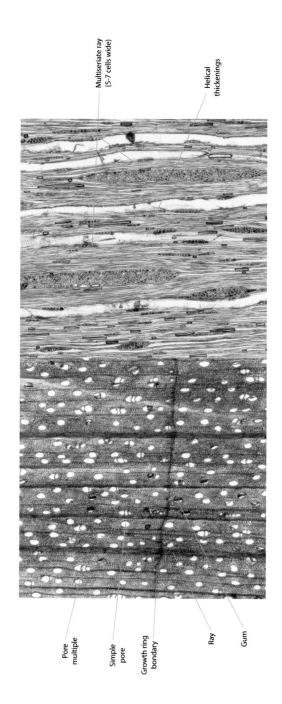

Pore multiple

Simple pore

Growth ring bondary

Ray

Gum

Multiseriate ray (5-7 cells wide)

Helical thickenings

Microscopic view of sugar maple. The transverse plane (*left*) shows simple and multiple pores. The tangential section (*right*) shows ray seriation of five to seven cells (5–7 rays cells of width), and helical thickenings, which are characteristic for the genus *Acer*. *Image courtesy Dr. Sarath Vega Gutierrez*

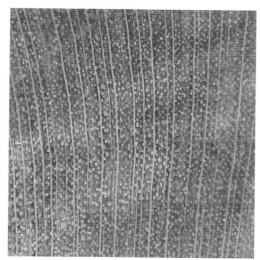

Transverse plane of sugar maple (*Acer saccharum*), showing a diffuse porous arrangement and two clear ray sizes. *Image courtesy Dr. Sarath Vega Gutierrez*

On the *transverse* plane (the side of the wood where you see the ends of the vessels and can count the growth rings), in temperate hardwoods, there are generally three arrangements of the vessels. They can occur in a *ring porous* arrangement, in which there is distinct earlywood and latewood without much transition, as defined by the size of the vessel (earlywood being bigger vessels with thinner cell walls, and latewood being smaller vessels with thicker cell walls); a *semi-ring porous* arrangement, where there is a gradual transition between large earlywood and small latewood; or a *diffuse porous* arrangement, where the earlywood and latewood look very much the same except at the growth ring boundary.

The pore arrangement of a hardwood can change, to a small degree, how the spalting develops. Woods with large earlywood (ring porous) provide a much more open avenue for fungi to travel. Pigment-type spalting fungi tend to do very well on ring porous hardwoods due to the more open structure.

13

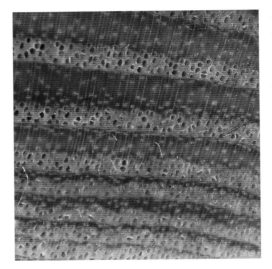

Transverse plane of black ash (*Fraxinus nigra*), showing a ring porous arrangement as well as vasicentric parenchyma around the latewood pores. *Image courtesy Dr. Sarath Vega Gutierrez*

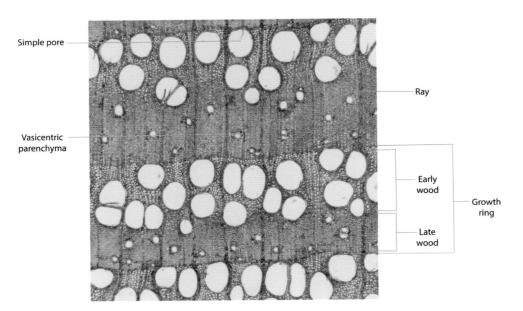

Microscopic view of black ash. *Image courtesy Dr. Sarath Vega Gutierrez*

Transverse plane of black walnut (*Juglans nigra*), showing a semi-ring porous arrangement with simple and multiple pores, and thin-banded parenchyma. *Image courtesy Dr. Sarath Vega Gutierrez*

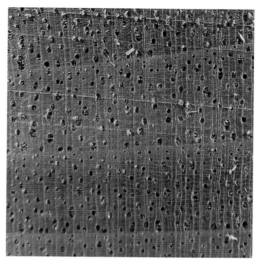

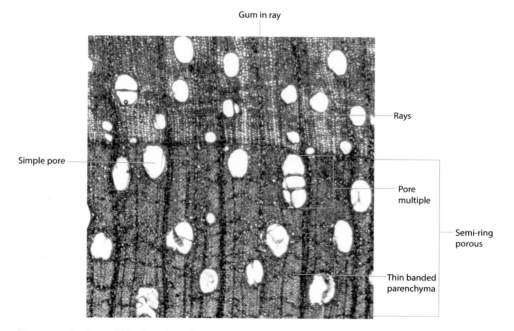

Microscopic view of black walnut. *Image courtesy Dr. Sarath Vega Gutierrez*

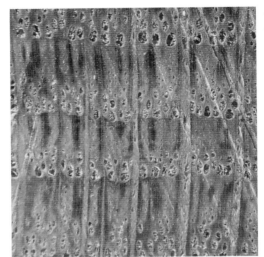

Transverse plane of white oak (*Quercus alba*). *Image courtesy Dr. Sarath Vega Gutierrez*

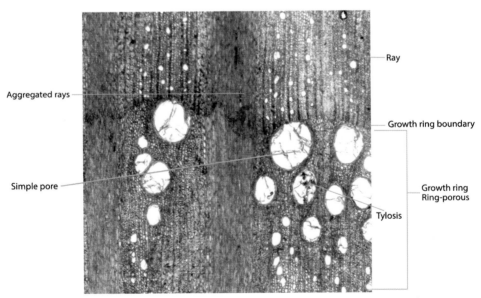

Ray

Aggregated rays

Growth ring boundary

Simple pore

Growth ring
Ring-porous

Tylosis

Microscopic view of white oak. The transverse plane shows simple pores, tyloses (the shiny stuff in the pores), and ring porous growth rings. *Image courtesy Dr. Sarath Vega Gutierrez*

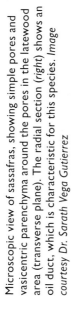

Transverse plane of sassafras (*Sassafras albidum*). *Image courtesy Dr. Sarath Vega Gutierrez*

Microscopic view of sassafras, showing simple pores and vasicentric parenchyma around the pores in the latewood area (transverse plane). The radial section (*right*) shows an oil duct, which is characteristic for this species. *Image courtesy Dr. Sarath Vega Gutierrez*

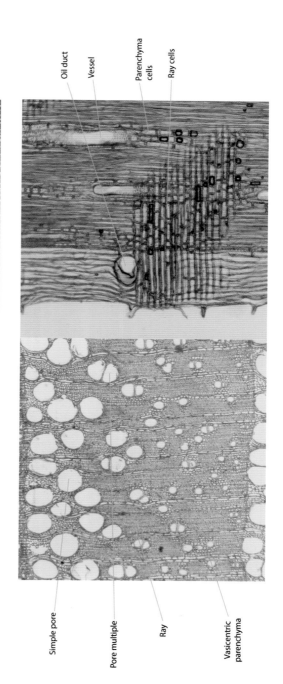

Oil duct

Vessel

Parenchyma cells

Ray cells

Simple pore

Pore multiple

Ray

Vasicentric parenchyma

While extractives—secondary metabolites produced by the tree to help defend against insects/bacteria/fungi/people—are less prevalent in temperate hardwoods, they do regularly occur, as do other types of cell blockage (occlusion). White oaks are easily distinguished from red oaks by their dendritic (long, vertical, wavy, "smoky" lines) latewood and abundant tyloses (that shiny stuff in the vessels in earlier photos). Many aromatic woods, such as sassafras, produce resins, gums, and/or oils as well as aromatic (smelly) extractives.

Extractives play a vital role in spalting. Since they are meant to help prevent movement of microorganisms and thereby limit decay, woods with a high extractive content (you can tell by their darker color, stronger smell, or stickier surface) spalt much, much more slowly than pale, odorless, white woods.

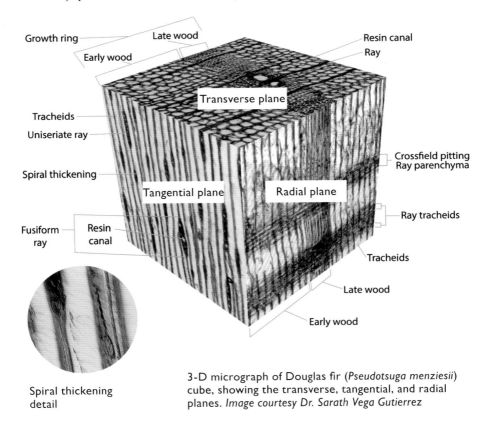

Spiral thickening detail

3-D micrograph of Douglas fir (*Pseudotsuga menziesii*) cube, showing the transverse, tangential, and radial planes. *Image courtesy Dr. Sarath Vega Gutierrez*

CONIFERS

Conifers are much more prevalent in cooler climates than warmer ones; hence this book will talk about conifers generally, instead of breaking them into temperate and tropical.

Conifers do not have vessel elements. Instead, the bulk of conifers is made up of longitudinal tracheids, which can be thought of like straws with pinched, tapered ends. Longitudinal tracheids function both for support and conduction, but unlike vessel elements, conduction happens primarily through the pits, which connect the tracheids side to side.

Conifers still have ray cells, both ray parenchyma and ray tracheids, though not all species have both types. Rays are generally unicellular (one cell wide) in conifers unless the ray contains a resin canal, in which case the ray expands around the canal. The cells that line the resin canals are called epithelial cells and secrete the resin or "pitch" we often associate with pine trees. Although any conifer can have resin canals, they occur with regularity only in Douglas fir, pines, spruce, and larch—a key ID characteristic.

Conifers, just like hardwoods, contain extractives. Some, such as cedrol from cedars, are aromatic, while some are more gumlike. All play a role in tree health, and many can affect humans. The extractives in conifers definitely affect spalting.

Western larch (*Larix occidentalis*) transverse plane, showing an abrupt transition between earlywood and latewood. *Image courtesy Dr. Sarath Vega Gutierrez*

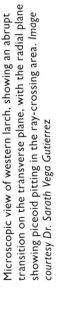

Microscopic view of western larch, showing an abrupt transition on the transverse plane, with the radial plane showing piceoid pitting in the ray-crossing area. *Image courtesy Dr. Sarath Vega Gutierrez*

Piceoid cross-field pitting

Abrupt transition

Growth ring

Late wood

Early wood

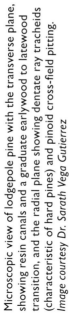

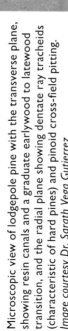

Lodgepole pine (*Pinus contorta*) transverse plane, showing resin canals and a gradual transition from latewood to earlywood. *Image courtesy Dr. Sarath Vega Gutierrez*

Microscopic view of lodgepole pine with the transverse plane, showing resin canals and a graduate earlywood to latewood transition, and the radial plane showing dentate ray tracheids (characteristic of hard pines) and pinoid cross-field pitting. *Image courtesy Dr. Sarath Vega Gutierrez*

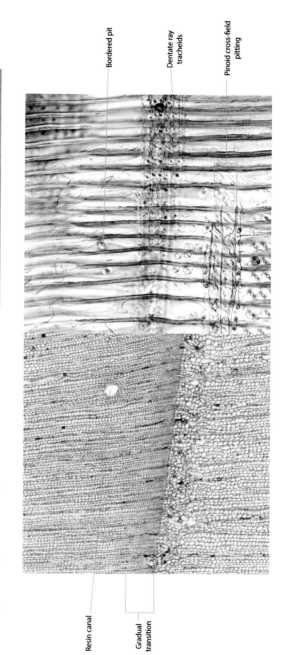

Bordered pit

Dentate ray tracheids

Pinoid cross-field pitting

Resin canal

Gradual transition

Conifers in temperate forests have growth rings, just like temperate hardwoods. Instead of being broken out by pore arrangement, however, they are generally broken down into two groups: those with abrupt transitions between the earlywood and latewood, and those with gradual transitions between the earlywood and latewood.

Because conifers don't have much variation in their cell types, a lot of ID relies on the type of pitting seen in their "cross field." As you can see in the images, when a ray cell (the cells running 90 degrees to the long axis of the tree) cross longitudinal tracheids, a sort of tic-tac-toe board is created. The shape of the pits within the crossfield is consistent within species and can help inform wood identification.

Conifers have a marginally different makeup (e.g., in terms of types of hemicellulose) than hardwoods, which is why rot fungi are generally either white rotting (enzymes slightly better suited for hardwoods) or brown rotting (enzymes slightly better suited for softwoods). Because white rotting fungi are the most likely to make zone lines, conifers seldom have zone lines form (although it does occur!). Pigment-type spalting, on the other hand, is usually done by soft rots, and such pigments occur with regularity on conifers. There will be more on the different types of decay in the next chapter.

TROPICAL HARDWOODS

Tropical hardwoods can be distinguished from temperate hardwoods through a collection of several factors: tropical hardwoods (1) tend not to have yearly annual rings and may have rings from drought cycles, rainy seasons, insects, etc., (2) tend to have a lot more parenchyma and/or can be best identified through their unique parenchymal arrangements, and (3) tend to have more extractives.

Many of the colored woods so loved by woodworkers and toymakers are tropical woods. Bloodwood and purpleheart, for instance, both have distinctive heartwood and colored extractives that turn the wood red or purple, respectively. There are temperate woods that are colored, such as osage orange, but these types of trees tend to be less frequent in temperate forests than tropical.

Because tropical woods tend to have more extractives, they tend to spalt much more slowly than their temperate counterparts. In general, DIY spalting using North American fungi on tropical hardwoods is a long, arduous process.

Bloodwood (*Brosimum rubescens*) transverse plane, showing the distinctive red color of the species. *Image courtesy Dr. Sarath Vega Gutierrez*

Microscopic view of bloodwood. Left side shows simple and multiple pores, tyloses, and aliform confluent parenchyma. The tangential section (*right*) shows gum ducts within the ray cells. *Image courtesy Dr. Sarath Vega Gutierrez*

Ray cells

Gum ducts

Tylosis

Ray

Simple pore

Aliform confluent parenchyma

Pore multiple

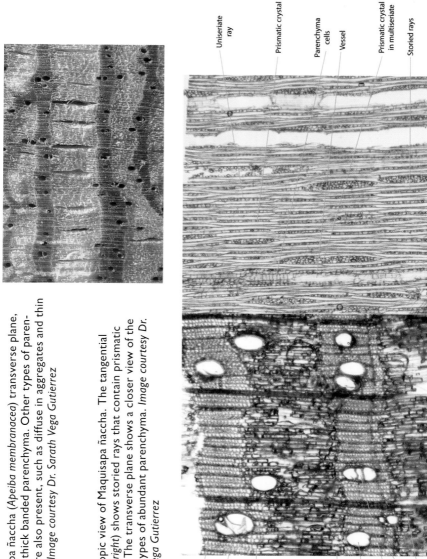

Maquisapa ñaccha (*Apeiba membranacea*) transverse plane, showing thick banded parenchyma. Other types of parenchyma are also present, such as diffuse in aggregates and thin banded. *Image courtesy Dr. Sarath Vega Gutierrez*

Microscopic view of Maquisapa ñaccha. The tangential section (*right*) shows storied rays that contain prismatic crystals. The transverse plane shows a closer view of the various types of abundant parenchyma. *Image courtesy Dr. Sarath Vega Gutierrez*

Uniseriate ray

Prismatic crystal

Parenchyma cells

Vessel

Prismatic crystal in multiseriate

Storied rays

Ray

Diffuse in aggregates parenchyma

Thick banded parenchyma

Simple pore

Pore multiple

Thin banded parenchyma

Air bubble

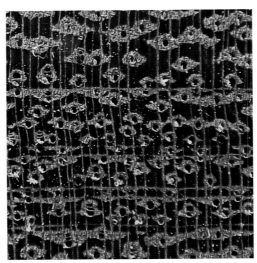

Purpleheart (*Peltogyne* sp.) transverse plane, showing the characteristic purple coloration that darkens to brown/black over time. *Image courtesy Dr. Sarath Vega Gutierrez*

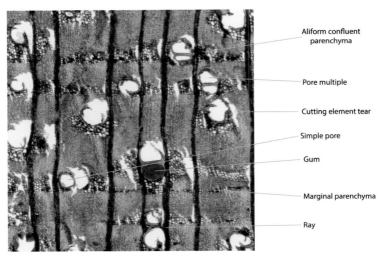

Aliform confluent parenchyma

Pore multiple

Cutting element tear

Simple pore

Gum

Marginal parenchyma

Ray

Microscopic view of purpleheart. Two types of parenchyma are visible: aliform confluent and marginal. The pores are simple and multiple, and some contain red gum. *Image courtesy Dr. Sarath Vega Gutierrez*

IN GENERAL

The bulk of movement in any tree is up and down, due to how the cells are oriented. Fungi may not necessarily follow this cellular highway, though. Especially in freshly felled timber, the easiest nutrients tend to be in the ray parenchyma. Although decay fungi (all spalting fungi are decay fungi, but not all decay fungi are spalting fungi) *can* digest the wood cell wall, most will preferentially digest easy sugars and carbohydrates first, which means *early-stage spalting tends to happen radially, not longitudinally.*

After the easy food has been removed, fungi may then run along the long axis as they forage and digest, but this is by no means a hard-and-fast rule. Tyloses, gum ducts, extractives, and other microbes can easily divert fungi, so while some *species* of fungi tend to grow in certain directions and patterns, generalizations cannot be made across the entire kingdom. Which, really, is part of the mystery, and fun, of spalting.

Chapter 2

Basics of Fungal Anatomy

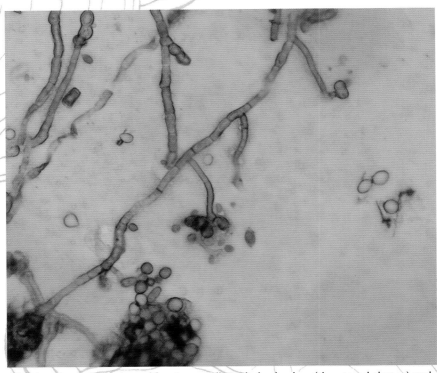

A basic, microscopic view of fungi generally includes hyphae (the strand shapes) and some sort of spore shape (for this fungus, the spores are round).

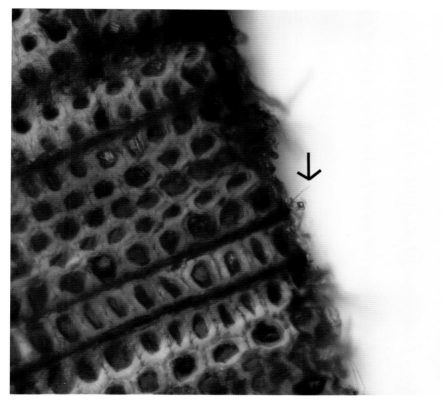

For size comparison, the same fungus exiting a conifer, transverse plane. The arrow points to a single hyphal strand.

INTRODUCTION

Fungi are so much more than the "spores" you hear everyone talking about. Fungi are filamentous (they grow in filaments, or strands) organisms with a chitinous cell wall (chitin is also the basic component in insect exoskeletons). Although once thought to be mostly plants, the fungal kingdom, we now know, is much more closely related to the animal kingdom than the plant kingdom. Fungi can have both sexual and asexual reproduction, and many species have more than two sexes.

Many fungi require oxygen, although not in great amounts. They grow both through elongation of their hyphae (the "strands") and through generation of spores, which can be thought of as the "fruit" of the fungus. Spores are small and light and easily dispersed through the air and water.

The Kingdom of Fungi is thought to contain several million species (between two and four million), and thus it can be hard to generalize across the entire kingdom. (In fact, you should *not* generalize across the kingdom, especially in terms of effects or toxicity. That would be like blaming dogs because a portion of spiders are venomous. Same kingdom, but VERY different parts.)

Three subgroups are of interest when discussing fungi: white rots, brown rots, and soft rots. Spalting fungi are defined as fungi capable of digesting wood and leaving *interior* color. This differentiates them from many mold fungi, which only color the surface of wood (and are often human irritants). Hence, spalting fungi are, at their heart, decay fungi, which are classified within the Dikarya subkingdom. White rots and brown rots are both in the division Basidiomycota, and the soft rots are in the division Ascomycota.

An extreme case of white rot (the brown is the original color of the wood) on maple heartwood

WHITE ROTS

White rot is a term used describe the damage done to wood by white rotting fungi (Basidiomycetes) when they degrade wood. These fungi are known for their degradation of lignin (one of the primary wood cell wall components: cellulose, hemicellulose, lignin), and it is the removal of the lignin—the structural support of wood, along with being slightly pigmented—that lightens and softens the wood. White rotted wood appears whiter than sound wood of the same species and is often very wet and mushy to the touch.

In its early stages, white rot can be a striking effect on wood, especially dark woods such as walnut and cherry. It can also be used to lighten wood before applying staining fungi, so that the colors can be seen. White rotting fungi primarily produce enzymes that break down hardwoods, although if hardwoods are scarce, they can do some damage to conifers.

 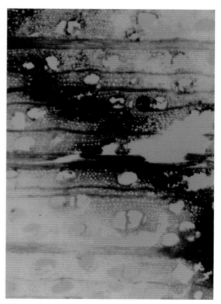

Black zone lines and light white rot on hand-carved *Casuarina equisetifolia* (common name: Avellano o pino australiano) wood. *Tahonga*, 2012, by Marco Antonio (Pepé) Tuki Hito. Easter Island, Chile. Contact information: pepetuki@hotmail.com

Microscopic view of a black zone line on sugar maple. *Photo courtesy Dr. Daniela Tudor*

31

Microscopic view of an orange zone line on pashako wood from the Peruvian Amazon, radial plane of the wood. The orange is primarily made up of an oily substance instead of melanin. *Photo courtesy Dr. Sarath Vega Gutierrez*

Macroscopic view of orange zone lines on pashako wood from the Peruvian Amazon

Zone lines on an unidentified species from the Peruvian Amazon. The zone lines appear black but are actually purple, which can be seen in the color diffusing from the main lines.

White rotting fungi are the primary producers of zone lines, the winding lines of color (often black or brown) that are readily associated with spalted wood. Zone lines are formed by fungi for a whole host of reasons, from genetic incompatibility ("this is my territory—get your own!") to lack of moisture (that thick melanin helps prevent wood water loss). Not all zone lines are composed of melanin, although it is a common component of white rotting fungal zone lines. Zone lines also come in a range of thickness and patterns, from fat melanin mats that look more like ink stains (common with *Xylaria polymorpha*—dead man's finger [ascomycete, not basidiomycete]), to thin lines that run in tandem (common with *Armillaria* spp.).

"White rot" or "bleaching" is considered one of the three main categories of spalting. Zone lines are a second.

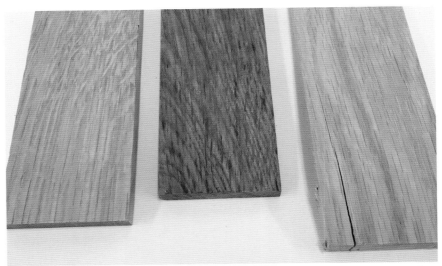

White oak (*left*), English brown oak (despite the name, this species is actually a white oak) (*center*), and red oak (*right*)

BROWN ROTS

Brown rot is the term used to described the dry, cubic, cracked wood left behind after brown rot fungi (Basidiomycetes) have colonized. Brown rot fungi do not generally degrade lignin but do degrade cellulose and hemicellulose, which leaves mostly the brown "structure" of the wood, without the filler. Brown rotting fungi produce enzymes primarily for conifer digestion and, as such, seldom colonize hardwoods—although there are many exceptions.

Brown rotted wood often crumbles at the touch and has relatively little use in spalting. A notable exception is English brown oak, which is a brown stain of oak trees caused by *Fistulina hepatica* (beefsteak fungus). English brown oak is known for its rich brown color and also its brittleness, which comes from the early stages of brown rot.

Both brown rots and soft rots are considered pigment-type spalting (they add pigment *to* the wood instead of creating color through removal, such as with white rot). Pigmentation is the third category of spalting.

Xylaria spp. (sporulating) covering a log in the Peruvian Amazon

A salmon-colored stain from an unidentified Ascomycete in the Peruvian Amazon

Downed log heavily colonized by *Chlorociboria* spp. in Chiloé, Chile. *Photo courtesy Patricia Vega Gutierrez*

SOFT ROTS

Soft rotting fungi are generally Ascomycete fungi (cup fungi) with marginal decay ability. The soft rotting fungi are characterized by their ability to selectively decay certain parts of the wood cell wall, causing cavities to form. They also tend to do well at higher moisture contents. Soft rots take far longer to visibly affect the wood than do white or brown rots, and do not have the same level of decay ability as the Basidiomycetes.

In terms of spalting, most of the pigment-type spalting fungi are ascomycetes. Some well-known examples are *Chlorociboria* species (elf's cup), which makes the distinctive blue-green spalt; *Xylaria polymorpha* (dead man's finger), which makes thick black zone lines; and the red/pink/blue stain produced by *Scytalidium cuboideum* (flaming dragon fungus).

Because many (but not all) soft rot fungi are very slow growing, they are often found on already well-decayed wood. Thus, even though soft rotting fungi alone take a very long time to destroy the working properties of wood, the wood they colonize is often so far down the decay path that it is already unusable. It is rare to find a piece of pigment-type spalted wood in the wild that is readily machinable.

The lab, of course, is an entirely different story.

Chapter 3

Debunking Myths

Kallfü Mamüll by David Leiva Leiva. The statue is a traditional Chilean sculpture from the Mapuche culture, called a *chemamüll* (wooden person). The sculpture was made from coigüe (*Nothofagus dombeyi*) and was heavily stained with the blue-green stain from *Chlorociboria* spp., as well as some orange zone lines from an unknown fungus, and made with a combination of traditional and modern carving. (h: 27 cm, w: 4–5 cm, t: 2–3 cm). *Photo courtesy David Leiva Leiva.* Website: www.casaetnica.cl

The number of urban legends that have cropped up around spalted wood in the United States is staggering. This shouldn't be surprising, since legends have always existed around spalted wood. A famous legend from Chiloé, Chile, for instance, tells the tale of the Trauco, a short, footless man who roams the forest and is irresistible to women—so much so that historically, unplanned pregnancies were attributed to woman-Trauco relations. The blue-green wood of *Chlorociboria* is very prevalent in Chiloé, though children and visitors are warned not to touch it, for it is "Trauco poo."

It is with some humor, then, that we can acknowledge the historical significance of spalting legends without giving them actual scientific weight. While the USA does not appear to have its own magical "Trauco poo wood," we do have endless recipes and conjecture on how to best produce spalted wood. This chapter will discuss and dissect some of the most popular recipes and then offer some more-reliable options for maximizing fungal pigmentation, while minimizing decay.

GROWING FUNGI VERSUS GROWING SPALTING

An entire book could be written on spalting recipes of the internet. Generally, the recipes are based on the idea that encouraging spalting shares something in common with growing edible fungi—and herein lies the major fallacy. Growing, say, shiitake mushrooms is about encouraging fungi to use resources and to fruit so that the fruiting bodies can be harvested. Even most laboratory testing is set up to promote decay, not spalting, by providing the best possible living environment for the fungi (so as to promote growth and mycelial mass).

This is the *opposite* of what we want for spalting fungi.

The trick to spalting is, quite literally, to promote stress in the target fungus. Since spalting is generally a stress response, whether to changes in climate or sensing a competing fungus, there is little incentive to produce the resource-intensive secondary metabolites (the pigments) if no danger is present. In fact, one of the reasons it is so difficult to keep spalting fungi pigmenting under laboratory conditions is that all of our standard fungal growth techniques were developed to keep fungi alive and *happy*, not alive and *stressed*.

The ideal spalted wood is heavily colored with minor decay. This can be accomplished only if the fungus spends some of its resources protecting its substrate, instead of focusing entirely on decay.

RECIPES

Most recipes for spalting wood involve ingredients that would promote fungal growth. These include things such as fertilizer (nitrogen is always a limiting factor, and fungi need it too), sugars (usually suggested in the form of beer, likely due to the malt), and mulch (for darkness). Some variations occur, such as ground-up reindeer antlers (I'm unsure what this provides), urination (urea is a great source of nitrogen), mayonnaise (I don't even know), oak leaves (nitrogen??? Some misplaced idea about tannins???), etc.

While many of these items are key for fungal growth, it's important to remember that in spalting, it's not the *growth* of the fungi that we are after. We want pigment production deep in the wood, with as little mycelium, and as little damage, as possible. It's also important to note that spalting usually requires at least a bit of decay, which means the fungus needs to start digesting the wood. As noted in chapter 1, fungi often utilize easy sugars in wood (ray cells) before attacking the more complex wood cell wall and breaking it down. Hence, providing easy sugar for the fungi (A) makes it "happy" and less likely to want to defend its territory and (B) slows down spalting, since the fungus will not generally begin wood decay until the easy sugar is gone.

CONDITIONS

So what *are* the best conditions for spalting? This is a really hard question to answer because no fungus is like another. Generalizations can't even be drawn across the same genus, and the wood species the fungus is growing on also plays a role.

In the lab, we consider "adequate" spalting to mean at least 20% of the wood surface *and interior* show some level of spalting (much more than that and it starts to get visually overwhelming). We use sugar maple (*Acer saccharum*) as our baseline. Sugar maple spalts to 20% spalting coverage with our baseline pairing of *Trametes versicolor* × *Polyporus brumalis* between eight and twelve weeks' incubation, where the wood is just above 30% moisture content (MC) and the temperate is held in the low to mid-80s with 80–90% relative humidity, incubated in the dark.

That's just one pairing though, on just one wood species. *Scytalidium cuboideum* takes between four and six weeks to *completely* stain the interior and exterior of sugar maple. Contrast that with *Chlorociboria* species, which takes six months or more to even hit the 20% marker on the same wood (and requires cooler conditions to do so).

There are, however, some generalizations that can be made. *In general:*
- *White rots* like warmer temperatures, between 78°F and 85°F.
- *Soft rots* like higher moisture contents, especially above 40% MC in the wood.
- *Lower MCs* can drive zone line production in such species as *Xylaria polymorpha*, which can use its melanized zone lines as a way to prevent moisture loss.
- *Zone lines* generally take longer to form than non-melanized pigments, with *Chlorociboria* an obvious exception.
- *Zone lines* generally require at least two fungi (*Xylaria* an obvious exception).
- *Ring porous woods* tend to pigment faster than diffuse porous woods, due to the larger vessels and ease with which the pigment can travel in the wood.
- *Light, white, non-decay-resistant woods* such as poplar and aspen are hard to generate zone lines on, since they are so quick to decay that fungi often just go after the wood and don't bother erecting barriers.
- *Conifers* do not readily get zone lines, since white-rot fungi tend not to colonize conifers.

Twelve weeks is a general average for how long something should take to spalt to 20% coverage, wherein live inoculum is applied to the cleaned wood, and the humidity and temperature are controlled (again, with *Chlorociboria* an obvious exception).

For more specifics on timelines, conditions, and pairings, the next few chapters break down each type of spalting and offer tips and tricks to maximize color while minimizing decay.

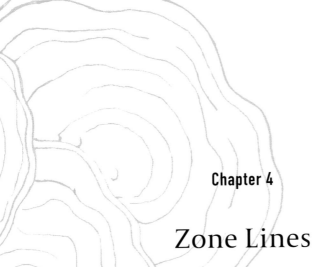

Chapter 4

Zone Lines

Although most people think of zone lines as "black line," they actually come in a huge variety of colors, thicknesses, and patterns. When attempting to induce zone line formation in wood, it is critical to first figure out what *type* of zone line you are after. What is your optimal width? What color? Do you want solitary lines, tandem lines, mats, or rosettes?

Once you know what kind of zone line you are after, you can select the appropriate fungi. Once you know what fungi you want, you can select your inoculation methodology. The following table provides an easy lookup for the most common zone lines and their fungi.

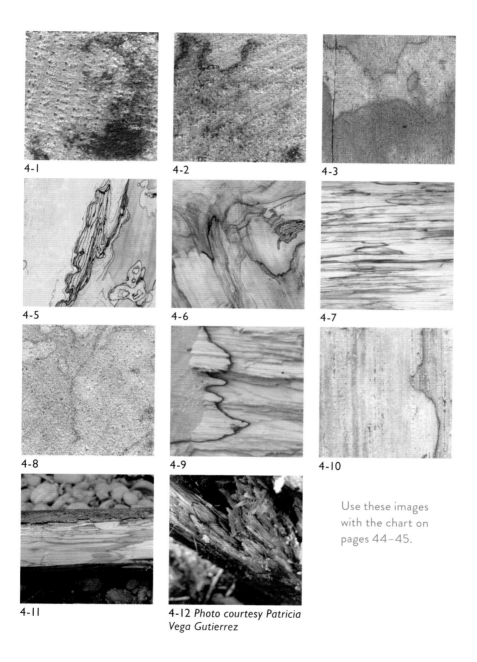

4-1

4-2

4-3

4-5

4-6

4-7

4-8

4-9

4-10

4-11

4-12 *Photo courtesy Patricia Vega Gutierrez*

Use these images with the chart on pages 44–45.

QUICK ZONE LINE LOOKUP CHART

Generalized across most temperate, low-to-medium-extractive woods. Chart is not exhaustive and shows only pairings known to work on sugar maple in laboratory settings.

Color	Thickness	Arrangement	Fungus A	Fungus B (if applicable)
black	>½"	mat	*Xylaria polymorpha*	N/A
	>¼"	solitary line	*Trametes versicolor*	*Bjerkandera adusta*
		solitary line	*Polyporus brumalis*	*Bjerkandera adusta*
	<¼"	tandem lines	*Armillaria* species	N/A
		rosette	Require several competing *Armillaria* species	
brown	>½"	line	*Fomes fomentarius*	N/A
	>¼"	line	*Inonotus hispidus*	N/A
yellow	>¼"	line	*Inonotus hispidus*	N/A
orange	variable	line	*Peniophora species*	N/A
red	variable	line	*Bjerkandera adusta*	*Polyporus brumalis*
			Polyporus brumalis	*Trametes versicolor*
pink	variable	line	*Trametes versicolor*	*Scytalidium cuboideum*
green	variable	line	*Trametes versicolor*	*Chlorociboria* species

Ideal Temp (°F)	Ideal MC	Incubation Time Required	Image (see page 43)
>75	<30%	>12 weeks	4-1
>80	>30%	>8 weeks	4-2
>80	>30%	>16 weeks	4-3
>80	unknown	>12 weeks	4-5
>80	unknown	>12 weeks	
>75	unknown	>6 weeks	4-6
>75	unknown	>10 weeks	4-7
>75	unknown	about 4 weeks (longer and it turns brown)	4-8
>80	>30%	unknown	4-9
>80	>30%	around 8 weeks (longer and the red bleaches away)	4-10
>80	>30%	>8 weeks	
>80	>30%	>6 weeks, but S. cuboideum must be inoculated first for 4–6 weeks, then T. versicolor added to the mix	4-11
<75	variable	around 2 years, with Chlorociboria inoculated first for 1.5 years, then T. versicolor added	4-12

METHODS

You can find methods for culturing fungi and copying fungal plates in chapter 8.

Rustic Method

Time frame: 1–2 years, depending on climate
Reliability: around 25%
Scientific knowledge required: none
Chance that your spouse/roommate will disapprove: low

This is by far the most common method used to generate zone lines. Take your wood, whether boards, logs, or firewood, and tightly pack it (the opposite of stickering). Make sure some wood is in ground contact. Leave uncovered and open to the elements. Come back in two years.

Why It Works: This is fundamentally just a firewood pile. It works because spalting fungi spores are in the air all the time, and in the soil. If you get lucky, the area you place your wood in will get the right combination of spores, and they'll land and begin growing and spalting. The best part is, once the fungal community is established, you can put wood in the same place over and over again and get approximately the same result. It's the community establishment that is tricky.

Quick and Dirty

Time frame: 1–2 years, depending on climate
Reliability: around 50%
Scientific knowledge required: none
Chance that your spouse/roommate will disapprove: low

This method is very similar to the Rustic Method, but you help establish the community. Stack your wood, making sure parts of it stay in ground contact. In places that are well aboveground (where on the wood doesn't matter so long as it is relatively clean), place known fungal cultures that give the type of spalting you want. With a bit of luck they will colonize the top part of your wood and then eventually compete with whatever is in your soil and air.

This isn't guaranteed, however; hence the low reliability.

Why It Works: Introducing known fungi can help those fungi establish before others, making sure they are in your wood to begin with. This can also help establish a site colony so you don't have to inoculate every batch, just the first one or two. There is still a chance that whatever is already in your wood/soil/air might outcompete the fungi you put on, so do keep that in mind.

The Warehouse

Time frame: 3–4 months
Reliability: around 75%
Scientific knowledge required: low to moderate
Chance that your spouse/roommate will disapprove: moderate, depending
how often they visit your storage shed

Like with the Rustic Method, tightly stack your wood. Unlike the previous two methods, do so somewhere where the wood is *not* in ground contact (see how we are slowly removing variables?) and is protected from rain and sun.

Make sure your wood is just *above* the moisture content ideal for your fungal pairings. If possible, control the heat of the space to at least 75°F, optimally 80–85. Inoculate your wood with known fungal cultures, ideally one petri plate every foot or so. Do not inoculate outside pieces of the pile. Let the entire thing grow for three to four months, preferably tarped or covered in some way to help with moisture loss, before checking.

To check the status of your spalting, remove one of the top or side pieces of wood and look at the next level. Resist the urge to disassemble the pile, since that will disturb the communities and increase incubation time. If you see black slicks across the surfaces, chances are good you have zone lines inside. If everything looks like regular wood, chances are your wood is too dry. Rehydrate and try again. If things look green, mold might have outcompeted your rot fungi. Leave the pile intact, wipe the mold off with 70% ethanol, and throw a couple of fresh plates of rot fungi in. Hope for the best.

You won't take the whole thing down at once. Remove layer by layer. If the layer you take down isn't spalted, wait another four weeks and come back. If the layer you take down *is* spalted, continue to the next layer. Remember that fungi do not grow evenly, so some layers may spalt while others may take longer. Be patient.

Why It Works: The more variables you can remove, the better. Getting rid of as many potential competitors from the soil and air goes a long way to making sure your fungi have their best chance of survival. Controlling temperature and humidity to some degree helps you make the wood ideal for your fungi, and not the ones that happen upon the wood while it's spalting.

The Scientific Warehouse

Time frame: 3–4 months
Reliability: around 95%
Scientific knowledge required: moderate
Chance that your spouse/roommate will disapprove: moderate to high

The primary difference between this method and the one before it is controlling the air. For this method, clean and cut your wood as much as possible, down to just above the dimensions you want (this is best suited to bowl blanks and small form wood over logs). Place fungal cultures on the broad faces of your wood, then tightly pack the wood into some kind of container—plastic tubs with snap-on lids work great for this. Ensure your wood is at the appropriate moisture content, then put the lid on and store the tub in a room where the temperature is in the correct range.

Ensure your wood stays within its correct moisture range with a moisture meter, but do not disturb the stacks. The more you move the wood around, the longer it will take.

Let sit for three to four months. To check, open the lid and assess the surface. If the surfaces are gray or black, or at least have some black streaking, chances are you have zone lines. If the surfaces are green, spray with 70% ethanol and put some fresh fungi down. If the surfaces are clear of fungi, your wood is too dry. Toss some tap water in there and reinoculate.

Why It Works: You've removed the moving-air component, which brings mold spores with it (and some decay ones as well). The only thing you aren't controlling at this stage is whatever is already in the wood. Given the correct temperature and humidity, there is nothing that should come in the way of good spalting with this procedure. Using less-than-ideal amounts of inoculum (ideal is a petri plate about every foot) may slow the spalting somewhat, but it won't stop it.

The best part is that plastic is porous. Once you've done a few rounds of spalting in your tub, you don't even have to inoculate your wood anymore! Just stick wet wood in the tub, put the lid on, and let it sit. The fungi will move back out of the plastic and into your wood! You've created your own portable ecosystem.

Tightly Controlled

Time frame: 6–12 weeks
Reliability: around 95%
Scientific knowledge required: high
Chance that your spouse/roommate will disapprove: low

If you are working with smaller pieces of wood for such things as marquetry or small turnings, you can remove the final variable of wood contamination. To do this, take your wood and either (A) microwave it until most of the water is removed (in blasts of about thirty seconds, giving two-minute sit time between, until the wood no longer feels wet after thirty seconds outside the microwave); (B) boil your wood (thirty minutes minimum at a rolling boil); or (C) autoclave (pressure cook) your wood for thirty minutes. The first two won't 100% sterilize your wood, but they'll take care of most of the surface molds, which is our major concern.

If you microwaved, be sure to add sterile water to your wood to get it back up to the right moisture content before inoculating. Whichever way you sterilize, immediately after place your wood in a plastic tub that you have cleaned *thoroughly* with 70% ethanol (that has since evaporated). Tightly stack your wood—the tighter the better—and cover. Wait one day. Go back the next day and, lifting the wood only as much as needed, place the fungal cultures. Make sure you are monitoring MC and temperature. Wait.

Why It Works: Removing the final variable, competing fungi, gives your spalting fungi time to grow as they please and interact wherever. You don't need as much inoculum because you don't need your fungi to outcompete anything. Getting rid of the competition also means you can, to some degree, control *where* your fungi grow, on the basis of where the petri plate mat lands.

This is by far the most reliable way to spalt wood and has a very high success rate. It fails only if the fungi used to inoculate the wood are contaminated, or if a contaminant gets in during the transfer from sterilization to tub. Occasionally it can take a bit longer than the Scientific Warehouse method, since the fungi, lacking antagonists, don't always interact right away, but the interactions you do get should be much more in line with your expectations.

Plastic bin filled with bowl blanks incubated with various fungi for twelve weeks. Green arrow points to a blank that looks "correct," in that black and white areas are present on the surface in large mats. Clear demarcations can be seen. The red arrow points to a blank that needs more time, since it shows evidence only of some light mold staining.

An important reminder that spalting does not occur evenly through a piece of wood, and where you place your inoculum matters. When left to nature, fungi tend to enter through the exposed end grain and colonize from either end of the log. This means that the ends are often more spalted than the middle. When you inoculate logs, you can inoculate more than just the end grain (through drilling holes, stripping bark, etc.), so that your spalting comes out a lot more even than what is pictured here. *Photos done with the help of Dr. Daniela Tudor*

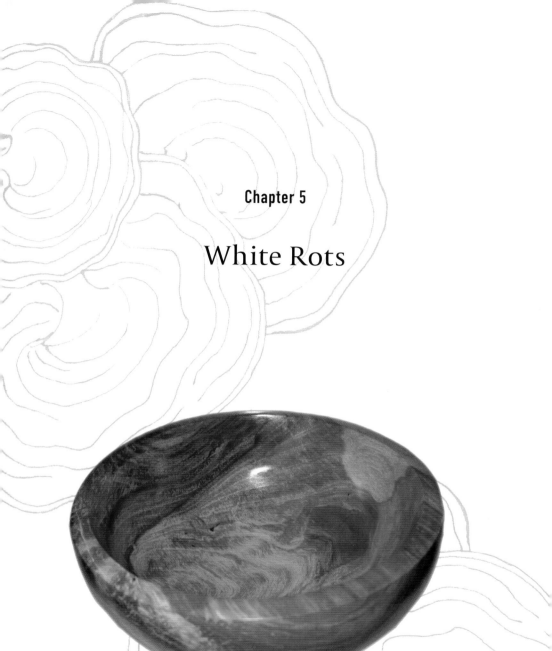

Chapter 5

White Rots

Moderate white rot on birch heartwood

CONTROL

Heavy white rot on maple heartwood

Although zone lines often come with white rot, white rot on its own can be a striking addition to wood. It is incredibly easy to white rot a piece of wood, but it is difficult to white rot it with *purpose*—to direct the rot where you want it to go. To do this, you need to tightly control the moisture content of your wood. Begin with dry wood and rehydrate only certain sections, whether by submerging half the wood in water (for no fungal growth under the water and then a gradient of growth up the piece as the water wicks) or dropping water onto just some areas of the wood.

Maple sapwood bowl. Left side was submerged in water, with a white rot fungus colonizing the "underwater" section as the piece was dried. The right side shows gray stain from *Scytalidium lignicola*, a blue-staining fungus.

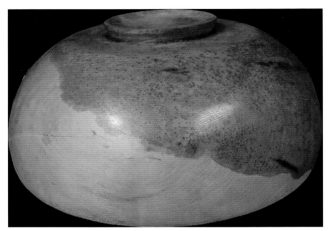

Maple heartwood bowl with white rot. The lighter portion was in contact with wet vermiculite, providing water for the fungi. The dark part was exposed and not in moisture contact, creating the distinct decay boundary.

Sugar maple sapwood bowl bathed in a low concentration of copper sulfate, then inoculated with *Trametes versicolor.* The texturing of the bowl is a result of the fungus attempting to sequester the copper and creating small islands of undecayed wood.

Since fungi generally need liquid water in wood to really do much, you can force the fungi to grow (generally) in certain areas by laying down a water path.

Other options for controlling white rot include the use of mild biocides, such as copper sulfate. In very low concentrations, copper sulfate will antagonize the fungus enough to sequester the copper—leaving islands of untouched wood in an otherwise ocean of white rot. Concentration of copper sulfate will vary by the porosity of the wood and the fungal species, so try a few out.

CONDITIONS

Most white rotting fungi like it warm and wet. Try to incubate wood for white rots at above 80°F (not above about 87°F though) and keep the relative humidity as high as you can. Keep your wood moisture content above 30%.

Be sure to keep a close eye on your white rot growth. This is the fastest type of spalting, and on some woods, such as birch, a day or two can be the difference between fantastic spalting and mush.

For methodology on inoculating white rotting fungi, follow the guides in chapter 4 for zone-lining fungi. Since most zone-lining fungi are also white rot fungi, their conditions tend to be the same.

COMMON WHITE ROT FUNGI

Below are some of the most common white rot fungi known to create reasonable white rot on wood, and that are relatively easy to acquire. While the list gives species, usually any species within that genus will work as well for general white rot.

- *Trametes versicolor*
- *Inonotus hispidus* (also causes a yellow stain)
- *Polyporus brumalis*
- *Pleurotus ostreatus*
- *Fomes fomentarius* (can also cause brown zone lines)

Sycamore heartwood with heavy colonization by *Inonotus hispidus*. Though the fungus is a white rot, it often leaves behind brownish zone lines and areas of exclusion.

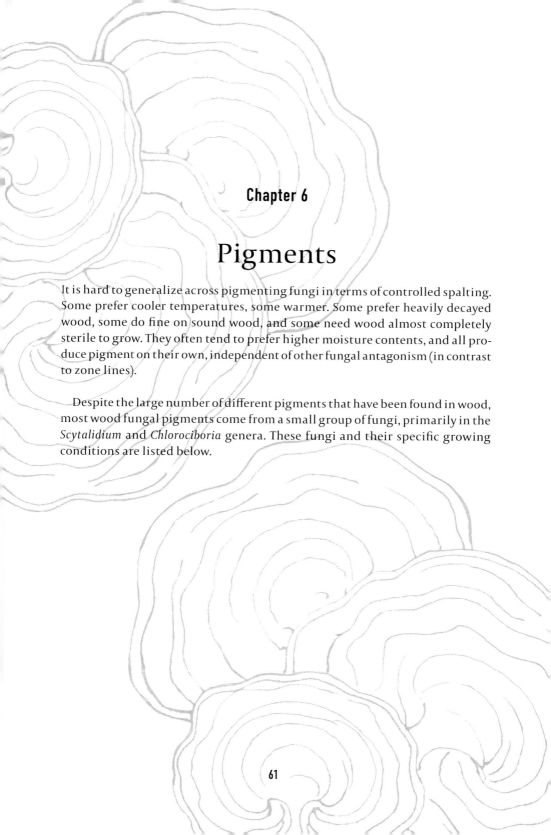

Chapter 6

Pigments

It is hard to generalize across pigmenting fungi in terms of controlled spalting. Some prefer cooler temperatures, some warmer. Some prefer heavily decayed wood, some do fine on sound wood, and some need wood almost completely sterile to grow. They often tend to prefer higher moisture contents, and all produce pigment on their own, independent of other fungal antagonism (in contrast to zone lines).

Despite the large number of different pigments that have been found in wood, most wood fungal pigments come from a small group of fungi, primarily in the *Scytalidium* and *Chlorociboria* genera. These fungi and their specific growing conditions are listed below.

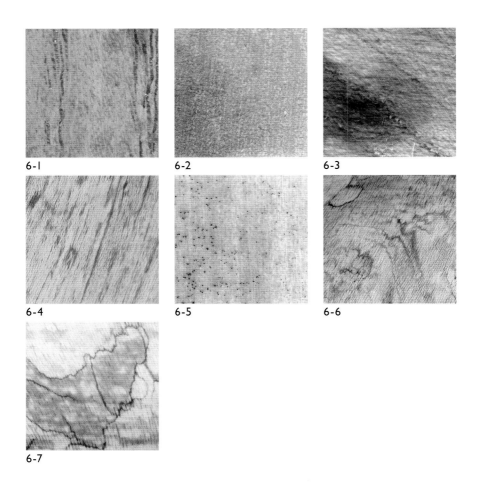

6-1

6-2

6-3

6-4

6-5

6-6

6-7

QUICK PIGMENT LOOKUP CHART

Generalized across most temperate, low-to-medium-extractive woods. Chart is not exhaustive and shows only fungi known to work on sugar maple in laboratory settings.

Color	Fungus	Ideal Temp (°F)	Ideal MC	Incubation Time Required	Better for Inoculation or Extraction?	Image (see page 62)
red	*Scytalidium cuboideum*	>75		4–8 weeks	either	6-1
blue		around 65–75		>8 weeks	inoculation	6-2
blue green	*Chlorociboria species*			>6 months	extraction	6-3
purple	*Scytalidium cuboideum*	>75	Unknown, but likely >35%	>16 weeks	inoculation—color comes from the overlap of the blue and red areas	6-4
				>36 weeks	extraction	6-5
yellow	*Scytalidium ganodermophthorum*	unknown		7–12 weeks	either	6-6
lime green				13–24 weeks	extraction	6-7
brown				25–36 weeks	either	6-8

METHODS

You can find methods for culturing fungi and copying fungal plates in appendix I.

Rustic Method

Time frame: 2–4 years, depending on climate
Reliability: around 5%
Scientific knowledge required: none
Chance that your spouse/roommate will disapprove: low

Unless you already have a very strong, robust community established, it is very difficult to induce pigment-type spalting by just leaving the wood alone. There are some exceptions, of course, such as areas with prevalent *Chlorociboria* (parts of Alaska, parts of Germany, most of Chiloé, Chile, etc.) or *Scytalidium cuboideum*. Over the years I've seen a dozen or so *Chlorociboria* communities robust enough to spalt a piece of wood just left on the ground in a short time frame. I have only *once* seen a *Scytalidium cuboideum* community. While the plot of land was incredible to behold—almost all the wood on it was bubblegum pink to hot-rod red—this is uncommon enough not to be worth your time.

Why It Doesn't Work: Most of the pigmenting fungi are very slow growing in the wild (in the lab is another story) and take years sometimes to establish just on one piece of wood. Some, such as *Chlorociboria*, tend to grow on already decayed wood, since they are not fantastic at wood decay (being soft rots, not white or brown rots). Hence, a pile of wood on the ground would likely decay to mush well before most pigmenting fungi even got started.

Quick and Dirty

Time frame: 2–4 years, depending on climate
Reliability: around 15%
Scientific knowledge equired: none
Chance that your spouse/roommate will disapprove: low

This method is very similar to the Rustic Method, but you help establish the community. Stack your wood, making sure parts of it stay in ground contact. In places that are well aboveground (where on the wood doesn't matter, so long as it is relatively clean), smear known fungal cultures that give the type of pigment you want.

Why It Often Doesn't Work: Most pigment fungi just can't outcompete mold fungi and Basidiomycete decay fungi in the time scale you likely want. Inoculating the top part of your wood might introduce the fungus, but whether or not it gets very far down before being outcompeted is a different story. Your wood *might* not have very aggressive fungi on it, or your soil might be relatively decay-fungus free, giving your pigmenting fungus a clear shot. Likely, however, this will not be the case, and your pigments will not establish until your wood is already fairly well decayed and no longer useful to you.

The Warehouse

Time frame: minimum 3 months, longer for *Chlorociboria* species
Reliability: around 35%
Scientific knowledge required: low to moderate
Chance that your spouse/roommate will disapprove: moderate, depending
how often they visit your storage shed

Like the Rustic Method, tightly stack your wood. Unlike the previous two methods, do so somewhere where the wood is *not* in ground contact (see how we are slowly removing variables?) and is protected from rain and sun.

Make sure your wood is just *above* the moisture content ideal for your fungal pairings. Heat the space as required. Inoculate your wood with known fungal cultures, ideally one petri plate every foot or so. Do not inoculate outside pieces of the pile. Let the entire thing grow for three to four months (longer for *Chlorociboria* spp.), preferably tarped or covered in some way to help with moisture loss, before checking.

To check the status of your spalting, slide under the tarp with a flashlight. If you see the color you're after, things are going well. If you don't, wait longer. It can take literally *years* for pigmenting fungi to color wood, and while this method removes a lot of the potential for decay fungi, there are still airborne fungi to compete with, as well as sound wood (not necessarily your fungus's ideal choice of substrate).

Why Results Are Variable: Your wood still harbors other fungi, as does the air. Inoculating this way gives your fungi a fighting chance, but there is still a lot of fighting to do!

The Scientific Warehouse

Time frame: minimum 3 months, fungus dependent
Reliability: around 75%
Scientific knowledge required: moderate
Chance that your spouse/roommate will disapprove: moderate to high

The primary difference between this method and the Warehouse Method is controlling the air. For this method, clean and cut your wood as much as possible, down to just above the dimensions you want (this is best suited to bowl blanks and small form wood over logs). Place fungal cultures on the broad faces of your wood, then tightly pack the wood into some kind of container—plastic tubs with snap-on lids work great for this. Ensure your wood is at the appropriate moisture content, then put the lid on and store the tub in a room where the temperature is in the correct range.

Ensure your wood stays within its correct moisture range with a moisture meter, but do not disturb the stacks. The more you move the wood around, the longer it will take.

Let incubate, based on the time needed for your fungus. To check, open the lid and assess the surface. If you see the color you want, you are on the right track! If you don't, toss some water in and wait longer. Most pigmenting fungi aren't great at decay, so there isn't as much danger of turning your wood to mush by allowing the wood to spalt too long as there is with zone-lining fungi.

Why It Works: You've removed the moving-air component, which brings mold spores with it (and some decay ones as well). The only thing you aren't controlling at this stage is whatever is already in the wood. Given the correct temperature and humidity, there is nothing that should come in the way of good spalting with this procedure. Using less-than-ideal amounts of inoculum (ideal is a petri plate about every foot) may slow the spalting somewhat, but it won't stop it.

The best part is that plastic is porous. Once you've done a few rounds of spalting in your tub, you don't even have to inoculate your wood anymore! Just stick wet wood in the tub, put the lid on, and let it sit. The fungi will move back out of the plastic and into your wood! You've created your own portable ecosystem. This works *especially* well for *Scytalidium cuboideum*. Once you have that fungus established in a tub, forever after will the wood you put in it turn red!

Fine powder from a handsaw, generated from a variety of pigmented wood. This dust can now be used for the extraction process.

Enough with the Nonsense

Time frame: 1–2 hours
Reliability: 100%
Scientific knowledge required: high
Chance that your spouse/roommate will disapprove: low as long as you are
outside

There is no reason to waste months or years encouraging pigment fungi to grow into wood. It is much more efficient to collect their pigments, either by buying them from a lab that collects them, or collecting them yourself, and then reapplying them to wood. This is especially true for *Chlorociboria*-stained wood, since it occurs readily in nature. Some of the other pigmenting fungi are harder to find and harder to be sure about in terms of their origin (don't just go sucking any red color from wood—some red-producing fungi are human pathogens), but buying the extracted pigment from a known vendor is a fast and reliable method to obtain spalting.

Poplar bowl spalted with the red pigment from *Scytalidium cuboideum*. Although the bowl was soaked with the pigment in solution, the pigment still moved primarily through the rays (just like the fungus would have), as seen in the pink radial streaks on the wood.

Remember—just because the pigment is extracted doesn't make the process any less "real." Spalting is just fungi excreting pigment into wood. Whether you let them do it on wood, or you get them to do it in a jar, or you collect it from one wood and put it on another, it's still literally the same process—fungal pigments moving through wood.

TO HARVEST PIGMENTS FROM NATURE

Collect a piece of pigment-type spalted wood from the forest. Use a handsaw to generate a powder from the wood (or any other pulverizing method you want). The smaller you can break the wood down, the more color you will get.

Take the powder/chunks and place them in a glass container. Canning jars work well for this. Start with just enough to cover the bottom of the jar. Pour your solvent of choice over the powder, usually about 100 mL, but enough so that the particles are floating maybe half an inch to an inch above the bottom of the jar. Gently swirl the solution. You should be able to see the color change almost immediately. Once you have your desired color, pour the contents of your jar into another, using coffee filter paper to strain the particulate. You should then have a brightly colored solution of fungal pigment and your solvent.

While DCM works better for the blue-green pigment from *Chlorociboria* species, the red pigment from *Scytalidium cuboideum* works equally well across a range of solvents, including DCM, acetone, chloroform, and pyridine.

Dichloromethane (DCM) is the best solvent. It holds a lot of color, extracts quickly, and evaporates quickly. It's also not good to use indoors or get on your skin (be sure to read its Safety Data Sheet before using), so if you use it, do so outside. You can also use acetone, but this solvent doesn't hold as much color, takes longer to absorb color from the wood, and takes longer to dry. It still works, but expect longer time frames across the board.

Once extracted and filtered, your solution can sit indefinitely until you are ready to use it. Note that some lower-quality acetones do interact with spalting pigments and may change the color moderately. DCM is quite stable.

TO USE PIGMENTS FROM A LAB

If you got your fungal pigment from a lab, then it likely came bound to a glass vial, with no powder or wood in the container. Spalting fungal pigments rarely if ever exist as a powder—their polarity drives them to bind to just about every-thing, which is why they are such great colorants.

To get the pigment off the glass, just add your solvent and recap. Let sit over-night *or* swirl the solution for long enough that the pigment moves back into the solvent. For acetone, this could take several hours, so best to let it sit over-night. For DCM, likely minutes.

Dry form of the pigment (known as "dramada") from *Scytalidium cuboideum*. The pigment is bound to the walls of the glass jar.

HOW TO USE THE EXTRACTED PIGMENTS

Through either method, you will end up with a colored solution in a solvent. You can use a glass pipette (no plastic—both acetone and DCM melt plastic) or a natural-hair paintbrush to pick up the solution and deposit it onto wood.

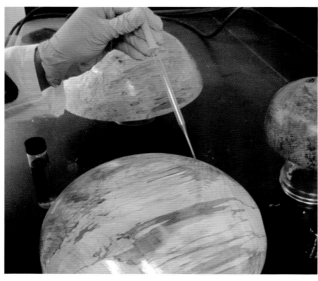

Spalting with extracted pigments. Use a glass pipette to suck up the pigment in solution, then slowly and carefully reapply to the wood.

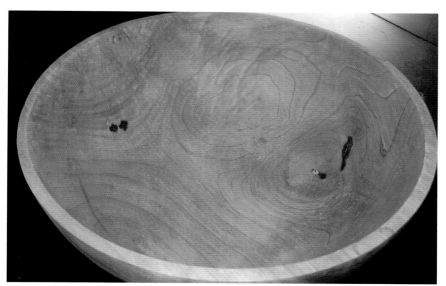

Magnolia wood, naturally a silver color, being spalted with the yellow pigment from *Scytalidium ganodermophthorum*. Two coats.

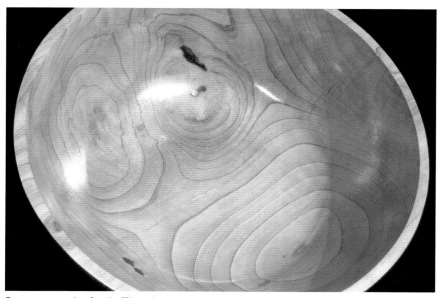

Five coats and a finish. The silver wood is now uniformly yellow.

The pigment tends to move the same way it would if a fungus were depositing it, if you put it on slowly. Remember, fungi deposit this pigment over *years*, and you're doing it in minutes. The structure of wood wasn't designed to handle large volumes. If you go slow, the pigment will move only marginally from where you placed it. If you "color in the lines," such as filling in a white-rotted area inside a zone line, unless you put on a ton of color, the pigment won't move past the line. Latewood, as well, can act as a natural boundary.

Bigleaf maple (*Acer macrophyllum*) with zone lines (induced) and green and yellow stain (from extracted pigments). The stains have stayed within the zone line boundaries.

Curly red maple (*Acer rubrum*) with zone lines (induced) and blue-green stain from *Chlorociboria* species (from extracted pigments). The change in color of the pigment comes from different numbers of layers. The bluer the color, the more layers were applied.

Knobcone pine (*Pinus attenuata*) spalted with red, blue-green, and yellow stain (extracted). Here the latewood was used as a natural boundary. Blue stain can also be seen on the bowl (the radial streaks), showing the spalting that occurred in the wood before it was taken to the lab.

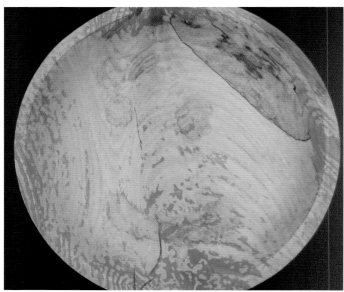

Spalting pigments are fairly "sticky." Even when being placed without boundaries, they tend to stay where you put them.

The most important thing to remember about using fungal pigments from an extract is that the pigments won't bind to the wood until the solvent has evaporated. This means that you need to let the wood dry between each coat, or the color won't build. In nature, these pigments get laid down layer after layer, and it is the increasing layers that build color (and change the color). For instance, the blue-green pigment from *Chlorociboria* starts off, generally, green and builds closer to blue with more coats. Number of coats is roughly equal to number of weeks incubation, so refer to the Quick Pigment Lookup Chart for color progressions.

Three different concentrations of the blue-green pigment from *Chlorociboria* species (known as "xylindein"). Left has the least amount of pigment. Right has the most amount of pigment.

Because the pigment must be put down in layers, dunking the wood, washing the wood, or pressure treating the wood (or a combination of these) doesn't help much. The color needs to be put on in thin layers and allowed to dry in between in order to build intensity.

Test blocks treated with the extracted pigment of *Scytalidium cuboideum* (red), *Chlorociboria* species (blue green), and *Scytalidium ganodermophthorum* (yellow). The colors are muted and do not progress past the surface, since they were applied via submersion, not dripping.

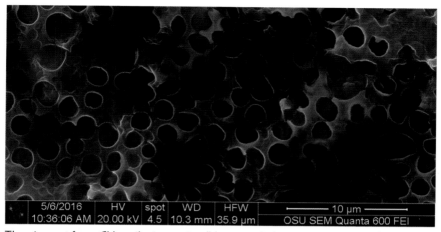

5/6/2016	HV	spot	WD	HFW	———— 10 µm ————
10:36:06 AM	20.00 kV	4.5	10.3 mm	35.9 µm	OSU SEM Quanta 600 FEI

The pigment from *Chlorociboria* species (blue green) deposits as a porous film. In order to build color, successive layers must cover existing gaps in the previous film. If a new layer is applied before the old one has set (before the solvent has evaporated), the new layer will simply form the same film shape as the previous (hence no change in color). *Photo courtesy Sarath Vega Gutierrez*

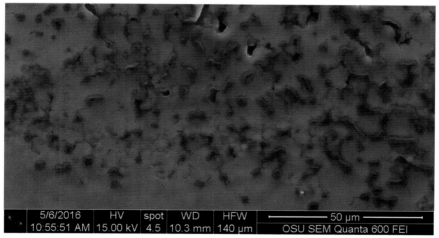

5/6/2016	HV	spot	WD	HFW	———— 50 µm ————
10:55:51 AM	15.00 kV	4.5	10.3 mm	140 µm	OSU SEM Quanta 600 FEI

The pigment from *Scytalidium ganodermophthorum* (yellow) also forms a porous film. In order to build color, successive layers must cover existing gaps in the previous film. If a new layer is applied before the old one has set, the new layer will simply form the same film shape as the previous (hence no change in color). *Photo courtesy Sarath Vega Gutierrez*

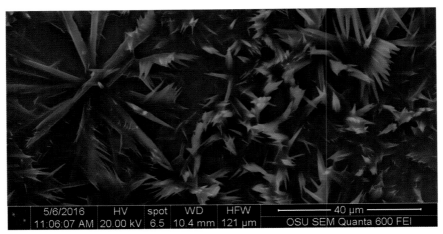

5/6/2016	HV	spot	WD	HFW	⊢————— 40 µm —————⊣
11:06:07 AM	20.00 kV	6.5	10.4 mm	121 µm	OSU SEM Quanta 600 FEI

The pigment from *Scytalidium cuboideum* (red) forms a flower-shaped crystalline structure. Although not much is known about the mechanics of its color change, it is theorized that the larger the crystals, the more toward the blue/purples the color moves. *Photo courtesy Sarath Vega Gutierrez*

The full range of pigment colors available currently from spalting fungi. *Image courtesy Patricia Vega Gutierrez*

Maple ornament by John Hyatt. The color comes entirely from extracted and reapplied xylindein from *Chlorociboria* species. *Photo courtesy John Hyatt, Sugar Land, Texas, USA.* Created after TXRX Labs, Houston, arranged for a demo on spalting by Dr. Robinson.

Curly maple pens (brass and ebony custom center band, gold tungsten nitride point and butt hardware, wire-burned fingertip grip detail, cyanoacrylate finish), showing the color change via different numbers of layers. Color comes entirely from extracted and reapplied xylindein from *Chlorociboria* species. *Photo courtesy John Hyatt, Sugar Land, Texas, USA.* Created after TXRX Labs, Houston, arranged for a demo on spalting by Dr. Robinson.

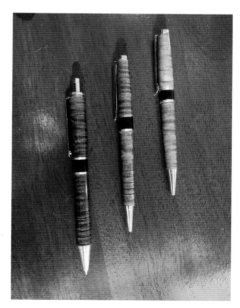

How coating and concentration change the color of extracted xylindein. While the percentage noted here is arbitrary, the effect of coating numbers and color of the base solution is striking. *Photo courtesy John Hyatt, Sugar Land, Texas, USA.* Created after TXRX Labs, Houston, arranged for a demo on spalting by Dr. Robinson.

Canvasback Love, by Joseph Scannell. The drawer pull is dyed blue green from extracted xylindein from *Chlorociboria* species. *Photo courtesy Joe Scannell*

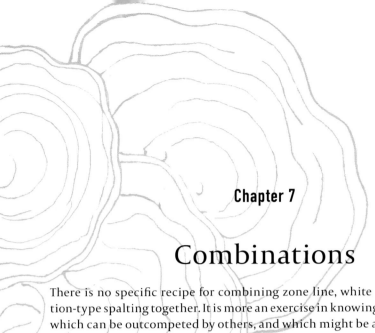

Chapter 7

Combinations

There is no specific recipe for combining zone line, white rot, and pigmentation-type spalting together. It is more an exercise in knowing which grow faster, which can be outcompeted by others, and which might be able to coexist.

If you choose only to live-inoculate, know that it will take years to effectively combine all the spalting types. Zone lines generally need to go on first, then the wood dried to kill the fungi. Then the pigments can be put on, one at a time, let grow, then killed before another pigment is placed. A pictorial guide is shown below.

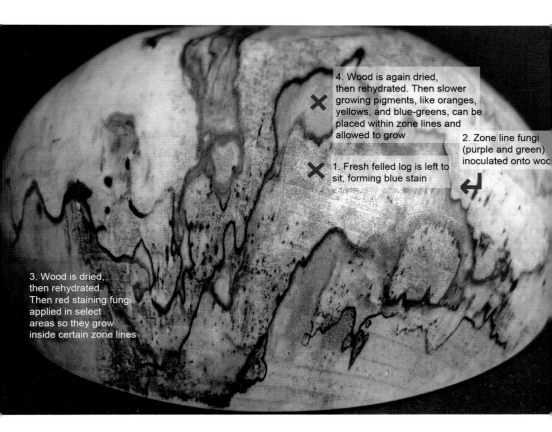

4. Wood is again dried, then rehydrated. Then slower growing pigments, like oranges, yellows, and blue-greens, can be placed within zone lines and allowed to grow

2. Zone line fungi (purple and green) inoculated onto woo

1. Fresh felled log is left to sit, forming blue stain

3. Wood is dried, then rehydrated. Then red staining fungi applied in select areas so they grow inside certain zone lines

If you choose to use the extracted pigments, combination spalting is quite easy. The zone lines must be grown on first (we do have extracted zone lines in the Applied Mycology Lab that can be applied with a pipette, like the extracted pigments, but they will not be available for home use anytime soon). Once the zone lines are to where you want them, machine the wood into its final form (and dry it). Sand to final sanding grit, *then* apply the extracted pigments. The solvent will raise the grain just a bit, but a light buffing with some 800-grit sandpaper will lay the grain right back down.

The difference between live inoculation and extracted inoculation is quite literally years. There's always a chance, too, for contamination in live inoculation bins. With extracted pigments you get exactly the color you want, exactly where you want it. Every time.

Chapter 8

Home Cultivation of Fungi

This pictorial guide will walk you through how to make media and then inoculate it with fungi from varying sources.
All photos done with the help of Dr. Daniela Tudor.

1 Be as clean as possible when working with fungal cultures. Soap and water is best for your hands, but for surfaces you can't run under water, 70% ethanol or 91% isopropyl will work as well.

2 You will need a stabilizing agent for your media. Pectin is readily available at most grocery stores but tends to liquefy easily.
   A better option is to go to a bulk food store and get gelatin or, even better, agar. Gelatin tends to be easier to find than agar, and cheaper; however, gelatin also tends to liquefy at room temperature, whereas agar stays solid. In the lab, we prefer agar. At home, use whatever is easiest.

3 You'll need a sugar base. Barley malt is one of the easiest (and cheapest) to come by.

4 Finally, you'll need something to hold the media in. You can buy premade petri plates on eBay or from lab supply companies, but it is easy to make your own. You'll be canning the media, however, so get a supply of glass canning jars, the smallest ones you can find.

5 Cook the media. Most spalting fungi are grown on 2% malt agar. Mix 1 L of water (distilled if possible), 20 g of malt, and 15 g of agar (substitute your stabilizer of choice, and feel free to play around with the amount if you aren't using agar) in a pot. Mix. Do not heat.

6 Pour or ladle the mixture into the canning jars, pouring in just enough that the bottom of the jar is covered. There is no need for a thick layer of media, unless you plan on storing the fungi for a very long time before use.

7 Place the jars in a pot. If you have a canning system, even better, since pressure helps with sterilization. Boil the jars for fifteen minutes, making sure the lid indents on the canning jars snap in. Once that happens, your jars are sealed and sterile.

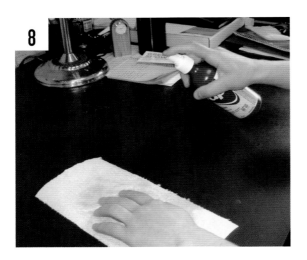

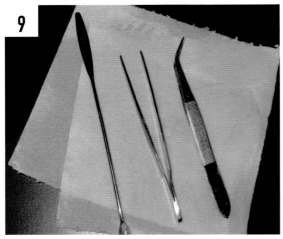

8 Let the jars cool for at least an hour. While you are waiting, clean any surface you will be working on.

9 Choose tools that can be easily washed/boiled and that have fine points. Small knives, spatulas, and tweezers are excellent choices. Be sure that all of these are as clean as you can get them.

10 You can take inoculum from anywhere, but there is always a risk of contamination. When trying to isolate from a fruiting body, crack the body open and take material from the inside. Place one small piece of the interior material into the center of your media. Keep in mind that fungi often colonize other fungi, so just because something grows on your plate doesn't mean it is the *correct* something.

You can also pull from spalted wood (**10b**), although the chance for contamination here is quite high since you have no idea what you are pulling out. To help, take from a freshly opened section of the wood; this should help contain airborne molds.

The best option is to take a scraping from a pure culture that you have made yourself or purchased from a reliable lab (**10c**).

11

12a

12b

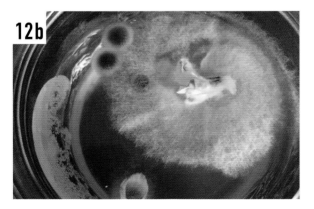

11 Whichever source you take from, make sure you place a very small section of material into the center of the media. This will help you determine later if you have contaminants. After the material is placed, put the lid back on and store the jar(s) in a clean space, like a drawer.

12 Check jars through the glass, not by opening them (as done in **12a**, for clarity). If a single colony has formed on the inoculation point, you have successfully cultured (to tell if you got the right fungus is a different story). After you have confirmed that you have the right fungus, you can proceed to use the media as inoculum, following the steps outlined in this book.

If there is more than one colony, especially if the colonies are different colors, you have contaminants (**12b**). Throw out the contents of the jar, clean the jar, and try again.

Chlorociboria (multipoint inoculation) growing on an agar medium in a canning jar. *Photo courtesy Mushroom Man of Peotone LLC: mush-roommanofpeotone.com*

Appendix I

Common Spalting Fungi and Where to Find/Buy Them

WHITE ROTS

Trametes versicolor (turkey tail) from a logging site near Hallojarvi (Gorynskoe) Lake near Lembolovo, 40 miles north from Saint Petersburg, Russia. Photo taken April 27, 2017. *Photo courtesy Alexey Sergeev: http://www.asergeev.com/index.htm*

Trametes versicolor (turkey tail) is a thin, shelf-shaped fungus with a broad worldwide distribution. It is edible, although the fruiting form is used mostly for alternative medicines. Although there are many fungi with similar shape, *Trametes versicolor* is often distinct due to its blue interior.

Polyporus brumalis (winter polypore) found near Orekhovo, north from Saint Petersburg, Russia, on October 22, 2016. *Photo courtesy Alexey Sergeev: http://www .asergeev.com/index.htm*

Polyporus brumalis (winter polypore) is a very "traditionally" shaped mushroom, with a stipe and cap. The cap is brown, but the white underside has pores, not gills, which is a major distinguishing feature. It also has a widespread distribution. It's not considered edible since it is so tough.

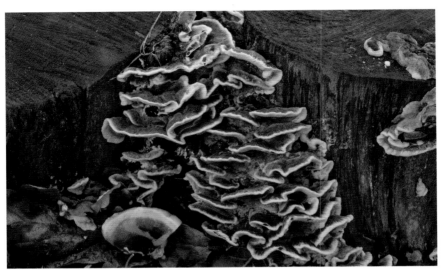

Bjerkandera adusta (smoky bracket) found near 4th South Pond on Elagin Island, Saint Petersburg, Russia, on October 7, 2017. *Photo courtesy Alexey Sergeev: http://www .asergeev.com/index.htm*

Bjerkandera adusta (smoky bracket) found near 4th South Pond on Elagin Island, Saint Petersburg, Russia, on October 7, 2017. *Photo courtesy Alexey Sergeev: http://www .asergeev.com/index.htm*

Bjerkandera adusta (smoky bracket) is widely distributed across Europe, North America, and Asia. It tends to grow on dead wood, as do most white rotting fungi. It has a number of look-alikes, so field identification needs to involve halving a fruiting body to see if there is a very strong color difference between the white "flesh" and the gray-black tubes. Lighter tubes separated by a distinct dark line would indicate *Bjerkandera fumosa*.

Pleurotus ostreatus (oyster mushroom) is grown commercially all over the world and, as such, is perhaps the easiest white rotting fungus to find. In the wild, one of the easiest ways to separate it from similar-looking oyster shapes growing from tree trunks is its smell—a distinctive bitter almond odor. While *P. ostreatus* is edible, it has several toxic look-alikes, so this one is best bought from a mushroom farm, or as a live culture.

An excellent reference for hunting white rotting spalting fungi (and from which much of the above information came), is *Fungi of Switzerland*, vol. 2, *Non-gilled Fungi*, by J. Breitenback and F. Kränzlin. While other guides may be smaller and have more "lay" language, the *Fungi of Switzerland* series gives detailed ID information as well as information about types of rot caused by the fungi. Many soft rotting fungi can be found in the book as well.

BROWN ROTS

Fruiting kown form of *Fistulina hepatica. Photo by Michael Kuo*

The only known reliable brown staining fungus is *Fistulina hepatica* (beefsteak fungus), known for causing the brown color in English brown oak. It is common in the UK but can be found in many places in the rest of Europe, North America, Australia, and some parts of Africa. It is edible (another common name is "poor man's beefsteak"). Like any brown rot, if left too long on wood it will eventually turn it unusable.

Fistulina hepatica is a bracket fungus that grows on tree trunks and is colored, as you might have guessed, a lot like raw beef.

SOLO ZONE LINERS

For fungi that require pairs to make zone lines, see the white rot section above.

Xylaria polymorpha (dead man's finger) both whole and halved (for visual reference), found near a rotting oak stump in Alexander Park, Pushkin (former Tsarskoe Selo), near Saint Petersburg, Russia, September 4, 2017. *Photo courtesy Alexey Sergeev:* http://www.asergeev.com/index.htm

Xylaria polymorpha (dead man's finger) is perhaps one of the most studied zone-line-producing fungi, with research into its melanin going back to the very early 1900s. It is a club-shaped Ascomycete fungus that often grows in the hollows of old stumps in North America, and directly on downed logs in South America. It has a worldwide distribution, though it is much more prevalent in some areas than others (such as the Amazon rainforest). The club forms can be blue early on but quickly darken to black as the season progresses. Touching the clubs will often leave a black residue on fingers. It is a soft rotting fungus, not a white rot, and, as such, takes longer to spalt wood than some of its more aggressive counterparts.

Fruiting form of *Inonotus hispidus. Photo by Darvin DeShazer*

Inonotus hispidus (shaggy bracket) is a white rotting Basidiomycete fungus that frequently grows on stressed, living trees. It has a distinct reddish top and yellow underside that often has yellow/orange "droplets" of condensation. It can be confused with *Fistulina hepatica* due to color and shape; however, *I. hispidus* has a hairy, fuzzy top while *F. hepatica* does not. *Inonotus hispidus* has a wide distribution across most of the world but is not common. It would be much easier to procure a pure culture from a culture bank than to find this fungus in the wild.

PIGMENTERS

Unfortunately, most pigmenting fungi don't make fruiting forms. Fortunately, the brilliant blue-green color comes from *Chlorociboria* species (elf's cup), which have a worldwide distribution. They're especially common in New Zealand and parts of Chile, especially Chiloé, but due to the distinct color of the wood can be easily found in most locations with sufficient rainfall.

The small blue-green cups tend to be more visible after a rain, but the stained wood is blue green year-round.

Chlorociboria aeruginosa (elf's cup) on rotting wood in Sosnovka Park, Saint Petersburg, Russia, August 14, 2017. *Photo courtesy Alexey Sergeev: http://www.asergeev.com/index.htm*

Close-up shot of *Chlorociboria* species in Michigan, USA

Log fully colored by *Chlorociboria* species in Chiloé, Chile. *Photo courtesy Patricia Vega Gutierrez*

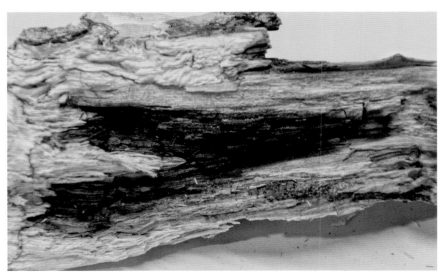

A more "usual" amount of *Chlorociboria* staining for the USA, also accompanied by some sparse black zone lines. *Photo courtesy Mich Williams*

WHERE TO BUY FUNGAL CULTURES

Although it is possible to collect fungi from the forest and do your own cultures, many people instead choose to purchase pure cultures from culture banks. This is especially necessary for many of the pigmenting fungi that do not make fruiting forms.

If you purchase from a culture bank, make sure they do not keep their fungi in cold storage. This often shuts down the pigmentation process, so while the fungi may still grow, they are often much slower at spalting. An ideal culture bank would keep spalting fungi actively growing, so that they are shipping in an aggressive state.

Some large culture banks include the American Type Culture Collection (ATCC), the University of Alberta Micro Herbarium (UAMH, and now housed at the University of Toronto), and the Madison (MAD) collection (at the US Forest Research Station in Madison, Wisconsin). The Applied Mycology Lab at Oregon State University has a spalting-specific culture bank as well, and cultures are never kept in cold storage.

There are many smaller outfits that sell spalting fungi as well. Note that if you are in the United States, most spalting fungi are considered plant pathogens and cannot be moved across state lines without a permit. If the place you are purchasing your fungi from does not require you to obtain a permit (or supply you one they have already obtained), please do not buy from them.

And, of course, spalting fungi can be purchased directly from the Applied Mycology Lab at Oregon State University, where most spalting research is conducted: www.northernspalting.com. The money from these sales goes directly back into more spalting research.

The most cost-effective way to purchase spalting fungi and spalting pigments, as well as to stay up to date on new research, is through the Spalting Cooperative (www.patreon.com/spalting). Fungi and pigments are available to cooperative members at significantly reduced rates, and members get to help direct current and future spalting research, as well as support the Applied Mycology Lab and its students.

Appendix II

Best Woods for Spalting

All wood can decay; therefore all wood can spalt. The *time frame* is what changes, since all spalting fungi are capable of wood decay but do so on very different schedules. Wood also contains many different types of extractives that vary by species (and vary in amount from tree to tree, even within different areas of the tree), which play a large role in decay resistance.

There are some general assumptions that can be made about wood, however.

Lighter woods spalt faster than heavier woods.
- Most weight in wood (aside from water) comes from extractives (think silica, for example).

Whiter woods spalt faster than darker woods.
- Most color in wood comes from extractives (think of walnut heartwood, osage orange, etc.).
- It takes longer for pigments to show on darker woods, making it appear as if the fungus needs more time to grow, when in reality there needs to be more pigment placed before it is visible.

Light, white woods do not zone-line well.
- Wood that has next to no decay resistance will pigment well but does not zone-line well, since the wood is too easily digested. Most fungi will just take what they want and leave, not bothering to mark their territory.

Tropical woods tend to have more extractives than temperate woods.
- This is a wide generalization, but in general, expect tropical woods to take longer to spalt than temperate woods, unless you are using tropical fungi in a tropical setting.

*Sugar maple (*Acer saccharum*) is an ideal spalting wood.*
- The wood spalts so well that it is often used as a control in spalting trials. If the sugar maple fails to spalt, something has gone wrong with the test. Interestingly, this does not apply across the board to all maples, *only* the hard maples. The types of sugars inside hard maples are theorized to play a role in this.

Dry wood doesn't spalt.
- Fungi need water to live.

Wood doesn't spalt underwater.
- (Most) fungi need oxygen to live.

Wood with strong odor doesn't spalt as well as wood without odor.
- Most smells in wood (think cedars) are aromatic extractives and play a role in decay resistance.

Hardwoods spalt better than softwoods for zone lines.
- Most zone-lining fungi are white rots, which tend to grow on hardwoods.

Softwoods and hardwoods spalt equally well with pigments.
- Most soft rotting pigmenting fungi grow equally well on hardwoods and softwoods.

Ring porous hardwoods pigment faster than diffuse porous hardwoods.
- The larger earlywood vessels make transport of pigments and fungi easier.

When just starting out, stick with medium-density white woods such as sugar maple and birch. Both spalt quickly and thoroughly and can give you good foundation of understanding before you move to more complex woods such as walnut, oaks, etc.

Notes on Woodworking with Spalted Wood

Spalted wood should not be worked with, or treated, like sound wood. Since spalting fungi are decay fungi, significant changes have occurred to the wood micro- and macrostructure during the spalting process (this obviously does not apply if the spalting occurred via extracted-pigment applications).

These are some general guidelines for working with spalted wood.

CHANGES IN STRENGTH AND DENSITY

Decay of any kind decreases wood density and strength. Do not use spalted wood for load-bearing applications. Consider it decorative use only unless stabilized with a resin.

CHANGES IN POROSITY

Decay makes holes in wood, though often these holes are too small to see. The holes affect strength *and* porosity. Expect finishes to sink much more readily into decayed areas than sound areas, causing uneven luster. Top-coat finishes (such as water-based polys) tend to be better for spalted wood, since they immediately form a surface layer and do not sink into wood. This avoids the differential-"shine" issues of oil-based finishes.

Expect to use more glue on spalted wood due to the increase in porosity.

CHANGES IN TEXTURE

Decay means the wood can more easily split apart, crumble, or flake when being pressed through a saw. Move spalted wood more slowly through machines and be prepared for tear-out.

Rotted areas will sand more quickly than the surrounding wood. Use foam sanding pads so you can vary pressure while sanding; otherwise the decayed areas will form divots in the surface of your wood.

CHANGES IN TONALITY

Due to the void space in the wood from decay, spalted wood has very different tonal qualities than sound wood. Spalted wood, especially wood affected by *Xylaria* species, has been used in the musical-instrument industry for centuries.

CHANGES IN CHEMICAL MAKEUP

The colors produced by spalting fungi are, fundamentally, chemicals. So are wood finishes. The spalting colors are known to interact with oils, and, as such, oil-based finishes can change the color of spalted wood (often muting the color or moving it to darker, less vivid shades) or cause the color to disappear entirely. Be aware of potential interactions between your finish and your spalting.

Notes on Woodturning with Spalted Wood

Image courtesy Eric Kawashima

Wood*turning* spalted wood presents a host of additional problems over and above what is listed in appendix IV.

The biggest issue encountered when turning spalted wood is tear-out—the ripping of fibers from punky areas by the tool. Many woodturners get around this by stabilizing their spalted wood with various resins, thus turning the wood into a wood-plastic composite (and evening out the density). There is nothing wrong with this method; indeed, there is a long history of woodturners using stabilized wood, but it also isn't *needed.*

Even the punkiest wood can be turned without significant tear-out or danger of the wood "blowing up" on the lathe. The trick to this is (1) *slow* speeds (well under 1,000 rpm, usually between 300 and 500 rpm), (2) don't be afraid to use a faceplate, and (3) riding the bevel on a gouge-style tool.

Left: A faceplate should always be used when roughing spalted wood. While the ones that come with the lathes will work, consider upgrading to a coated faceplate with more than four screw holes. *Image courtesy Oregon State University*

Right: Always run the tailstock up when roughing out spalted wood, even if the point only partially connects. The more points of contact, the better. *Image courtesy Eric Kawashima*

Although much maligned in modern woodturning, faceplates are excellent insurance when turning spalted wood. The multiple points of contact mean more stability—something that is sorely needed when the density changes in the wood mean that there will always be some vibration. Cutting insets for expansion chucks often leads to cracked wood since the spalted wood cannot hold the pressure, and so the chucks often crack or pull off completely. Good-quality faceplates—those with a coating and more than four holes—are ideal for holding wood on the lathe. If you are turning a piece that will not be completely rounded, consider using #8 wood screws to start with until the piece is roughed. After roughing, if the #8s have wiggle to them, remove them and put in #10s. This doesn't involve removing the faceplate and will keep everything centered.

As an important reminder: the more points of contact, the better. Even when using a faceplate, run your tailstock up if possible.

Roughing cut on a piece of spalted wood by using a traditionally ground bowl gouge. *Image courtesy Oregon State University*

Woodturning during the past ten years has made a move toward scrapers as a general all-purpose tool, due to the ease of use, lack of sharpening in the "disposable" lines, and relative safety of the tools. Unfortunately, scrapers are not ideal for many softer woods and wreak havoc with spalted wood. The decay in spalted wood equates to small, microscopic holes and fissures in the wood, which means that separation of cells happens much more readily. Tear-out is nearly inevitable with most turning tools that scrape, since the force being exerted on the wood is pulling the cells, which then break off from the surrounding tissue and pull out.

Gouges, especially traditionally ground bowl gouges (no fingernail profile, but more point than a bottom feeder), can be used in such a way that a fine, light cut is taken (like a finishing cut), wherein the bevel is pressed gently against the bowl during the cutting action. This secures the cells down before the cut begins, minimizing tear-out as the cells are supported. Think of it as the difference between ripping off a Band-Aid (scraping) versus pulling your skin taut and *then* ripping the Band-Aid off. The latter is less painful and tends to remove far less skin because of the support provided.

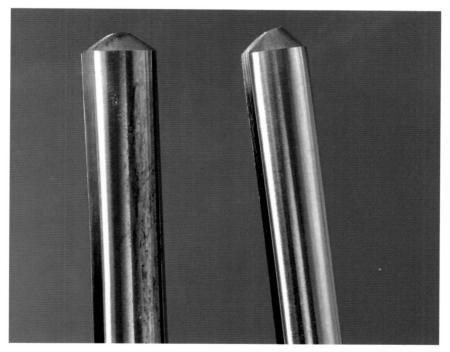

Right: Factory grind on a traditional bowl gouge (Robert Sorby brand).

Left: Slightly modified grind for helping with interior transition cutting on spalted wood. This grind is in between a bottom feeder grind and the factory grind.

Appendix V

Most-Common Questions and Their Answers (FAQ)

What is spalting?
Spalting is any color in wood produced by a fungus. It is not color produced by the tree (such as in box elder, lilac, etc.), color produced by minerals, chemical interactions, etc. Spalting can be zone lines, pigment, white rot (bleaching), or any combination thereof. It encompasses spalted wood found in nature and that produced in a garage or lab.

What fungi can spalt wood?
A fungus must be able to degrade wood in order to spalt. Hence, only decay fungi are spalting fungi, meaning general airborne molds, though some do produce surface pigment, are *not* spalting fungi. White rots, brown rots, and soft rots are all capable of spalting, though not all produce pigment along with decay.

What woods can spalt?
All wood can decay; therefore all wood can spalt. Woods with more decay resistance (more extractives) take longer to spalt. Woods with very little decay resistance may not spalt as well, due to decaying very rapidly.

How long does spalting take?
No concise answer can be given for this. See the chapters in this book for more information about why.

Is spalting dangerous to people?
No. Spalting is not mold (and in mold's defense, most molds are completely benign too).

Can you spalt dry wood?
Yes, if you get it wet enough.

Do additives help spalting go faster, such as sugars, beers, urine, mayonnaise, leaves, etc.?
No. They make it take longer.

What type of barley malt is best for spalting?
Laboratory grade is best. Barring that, barley malt extract tends to work the best.

Can I grow spalting fungi on liquid cultures or liquefy the culture so it can be sprayed on the wood?
You *can*, but it isn't recommended. Diluting the food base, spreading out the inoculum, or increasing the food base all changes the dynamic of colonization. Petri plates with malt agar are ideal in that they have just the right amount of food and growth to help a fungus establish. Spread out the inoculum, increase or decrease the food, and your fungus has less of a chance of establishing.

Is spalted wood dangerous to woodturn?
Not if you do it slowly, with sharp tools, and ride the bevel. See appendix IV for more information.

Is spalted wood food safe?
As far as is currently understood, yes. However, spalted wood is much more porous than sound wood, so don't expect your spalted bowl to hold soup.

Appendix VI

Spalting Appearance by Wood Species

This information was first published in *Spalted Wood: The History, Science, and Art of a Unique Material* (Schiffer Publishing). Additional lookup plates can be found in that book. The most pertinent ones are reprinted here.

COMPARATIVE GUIDE TO SPALTING

The following pages show spalting on a variety of wood species from around the world. Some are spalted with known fungi, and others were simply found. All are on unfinished wood, so the true effect of the spalting on each wood species can be seen. All are at twelve weeks' incubation for comparison.

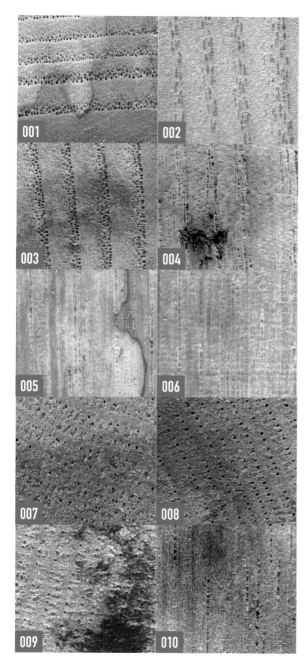

Pacific Northwest
wood species of North
America

Ash (*Fraxinus latifolia*)
001 White rot and zone
lines resulting from the
combination of *Trametes
versicolor* and *Polyporus
brumalis*

002 Pink stain from
Scytalidium cuboideum

003 Blue stain from
Scytalidium lignicola

004 Green stain from
Chlorociboria aeruginascens

Chinkapin (*Chrysolepis
chrysophylla*)
005 White rot and zone
lines resulting from the
combination of *Trametes
versicolor* and *Polyporus
brumalis*

006 Pink stain from
Scytalidium cuboideum

007 Yellow stain from
*Scytalidium
ganodermophthorum*

008 Blue-and-yellow
stain from *Scytalidium
lignicola*

009 Zone lines from
Xylaria polymorpha

010 Green stain from
Chlorociboria spp

Cottonwood (*Populus trichocarpa*)

011 White rot and zone lines resulting from the combination of *Trametes versicolor* and *Polyporus brumalis*

012 Pink stain from *Scytalidium cuboideum*

013 Yellow stain from *Scytalidium ganodermophthorum*

014 Blue stain from *Scytalidium lignicola*

015 Zone lines from *Xylaria polymorpha*

016 Green stain from *Chlorociboria aeruginascens*

Dogwood *(Cornus nuttallii)*

017 White rot and zone lines resulting from the combination of *Trametes versicolor* and *Polyporus brumalis*

018 Pink stain from *Scytalidium cuboideum*

019 Yellow stain from *Scytalidium ganodermophthorum*

020 Yellow stain from *Scytalidium lignicola*

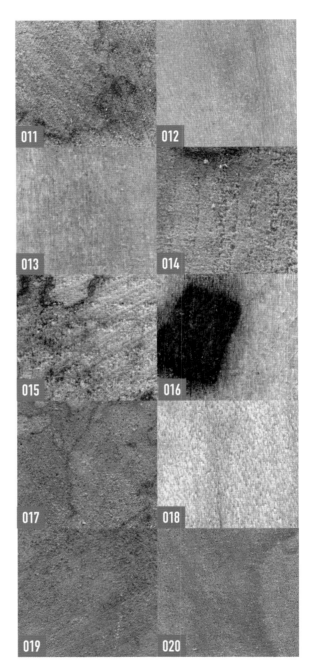

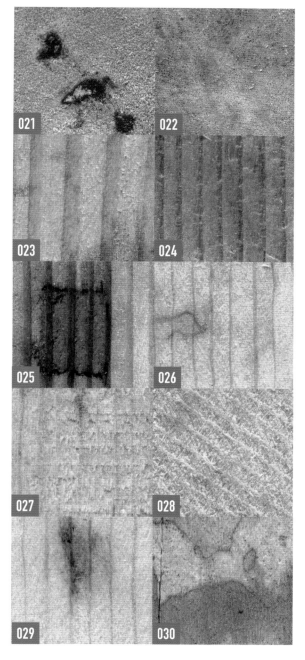

021 Zone lines from
Xylaria polymorpha

022 Green stain from
Chlorociboria aeruginascens

Douglas fir (*Pseudotsuga menziesii*)
023 Pink stain from
Scytalidium cuboideum

024 Blue stain from
Scytalidium lignicola

025 Green stain from
Chlorociboria aeruginascens

Lodgepole pine (*Pinus contorta*)
026 White rot and zone lines resulting from the combination of *Trametes versicolor* and *Polyporus brumalis*

027 Pink stain from
Scytalidium cuboideum

028 Yellow stain from
Scytalidium ganodermophthorum

029 Green stain from
Chlorociboria aeruginascens

Madrone (*Arbutus menziesii*)
030 White rot and zone lines resulting from the combination of *Trametes versicolor* and *Polyporus brumalis*

Mountain hemlock
(*Tsuga mertensiana*)
031 Pink stain from
Scytalidium cuboideum

032 Yellow stain from
*Scytalidium
ganodermophthorum*

033 Blue stain from
Scytalidium lignicola

034 Green stain from
Chlorociboria aeruginascens

Myrtle (*Myrtus* sp.)
035 White rot and zone
lines resulting from the
combination of *Trametes
versicolor* and *Polyporus
brumalis*

036 Pink stain from
Scytalidium cuboideum

037 Blue stain from
Scytalidium lignicola

038 Zone lines from
Xylaria polymorpha

039 Green stain (appear-
ing brown) from
Chlorociboria aeruginascens

Noble fir (*Abies procera*)
040 Pink stain from
Scytalidium cuboideum

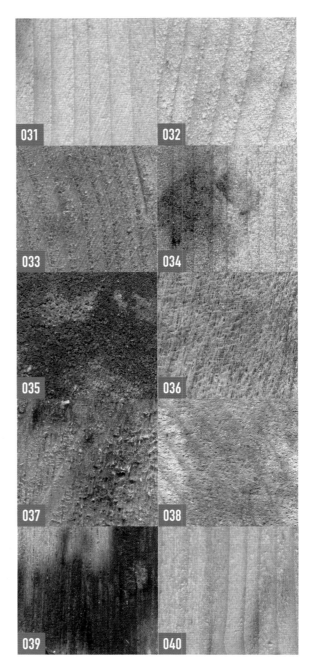

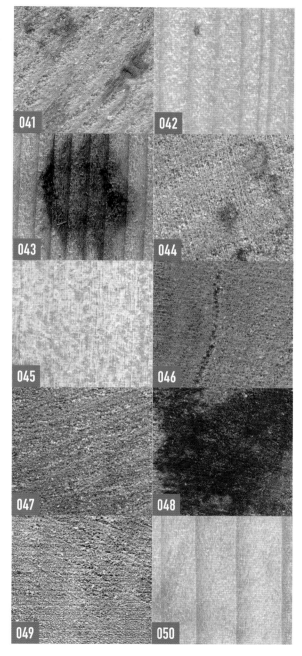

041 Yellow-and-blue stain from *Scytalidium lignicola*

042 Zone lines from *Xylaria polymorpha*

043 Green stain from *Chlorociboria aeruginascens*

Oregon maple (*Acer macrophyllum*)
044 White rot and zone lines resulting from the combination of *Trametes versicolor* and *Polyporus brumalis*

045 Pink stain from *Scytalidium cuboideum*

046 Yellow stain from *Scytalidium ganodermophthorum*

047 Blue stain from *Scytalidium lignicola*

048 Zone lines from *Xylaria polymorpha*

049 Green stain from *Chlorociboria aeruginascens*

Pacific silver fir (*Abies amabilis*)
050 Pink stain from *Scytalidium cuboideum*

051 Blue stain from
Scytalidium lignicola

052 Green stain from
Chlorociboria aeruginascens

Pacific yew (*Taxus brevifolia*)
053 White rot resulting
from the combination of
Trametes versicolor and
Polyporus brumalis

Port Orford cedar
(*Chamaecyparis lawsoniana*)
054 Pink stain from
Scytalidium cuboideum

055 Blue stain from
Scytalidium lignicola

056 Green stain from
Chlorociboria aeruginascens

Red alder (*Alnus rubra*)
057 White rot and zone
lines resulting from the
combination of *Trametes
versicolor* and *Polyporus
brumalis*

058 Pink stain from
Scytalidium cuboideum

059 Yellow stain from
*Scytalidium
ganodermophthorum*

060 Green stain from
Chlorociboria aeruginascens

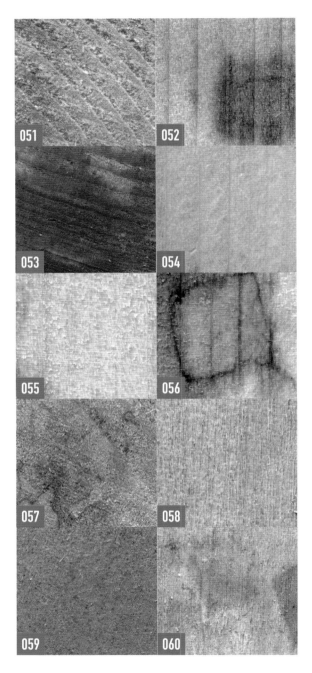

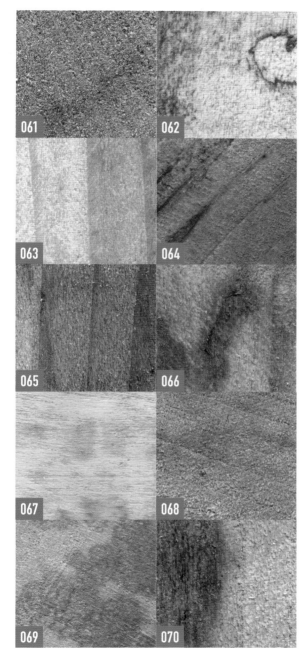

061 Blue stain from *Scytalidium lignicola*

062 Zone lines from *Xylaria polymorpha*

Redwood (*Sequoia sempervirens*)
063 Pink-and-blue stain from *Scytalidium cuboideum*

064 Zone lines from *Xylaria polymorpha*

065 Green stain from *Chlorociboria aeruginascens*

Sweet cherry (*Prunus avium*)
066 White rot and zone lines resulting from the combination of *Trametes versicolor* and *Polyporus brumalis*

067 Pink-and-blue stain from *Scytalidium cuboideum*

068 Yellow stain from *Scytalidium ganodermophthorum*

069 Zone lines from *Xylaria polymorpha*

070 Green stain from *Chlorociboria aeruginascens*

Walnut (*Juglans hindsii*)
071 White rot and zone lines resulting from the combination of *Trametes versicolor* and *Polyporus brumalis*

072 Pink-and-blue stain from *Scytalidium cuboideum*

073 Yellow stain from *Scytalidium ganodermophthorum*

074 Green stain from *Chlorociboria aeruginascens*

Western larch (*Larix occidentalis*)
075 White rot and zone lines resulting from the combination of *Trametes versicolor* and *Polyporus brumalis*

076 Pink stain from *Scytalidium cuboideum*

077 Green stain from *Chlorociboria aeruginascens*

Western red cedar (*Thuja plicata*)
078 Pink stain from *Scytalidium cuboideum*

079 White rot and zone lines resulting from the combination of *Trametes versicolor* and *Polyporus brumalis*

080 Green stain from *Chlorociboria aeruginascens*
White oak (*Quercus*

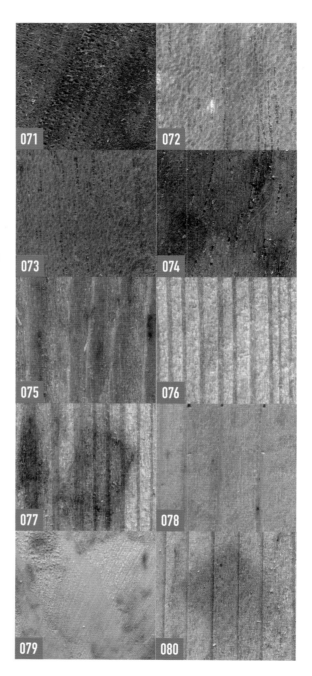

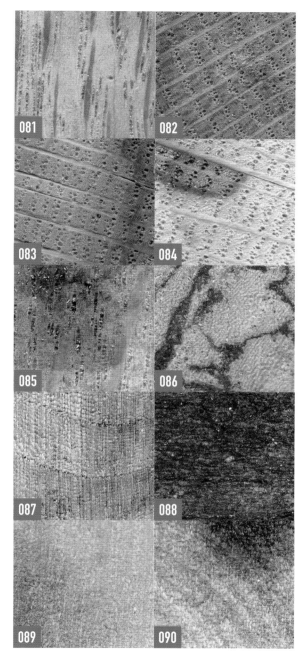

garryana)
081 Pink stain from *Scytalidium cuboideum*

082 Yellow stain from *Scytalidium ganodermophthorum*

083 Blue stain from *Scytalidium lignicola*

084 Zone lines from *Xylaria polymorpha*

085 Green stain from *Chlorociboria aeruginascens*

Northeastern wood species, including common ornamentals, of North America

American elm (*Ulmus americana*)
086 Yellow stain and zone lines of *Inonotus hispidus*

087 Blue stain of *Ophiostoma piceae*

088 Zone lines of *Xylaria polymorpha*

Aspen, quaking (*Populus tremuloides*)
089 Pink-and-blue stain from *Scytalidium cuboideum*

090 Blue stain from *Scytalidium lignicola*

091 Green stain from
Chlorociboria aeruginascens

092 Pink stain from
Monascus pilosus

093 Pink stain from
Monascus ruber

094 Pink stain from
Scytalidium cuboideum

095 Yellow stain from
*Scytalidium
ganodermophthorum*

096 Zone lines from
Xylaria polymorpha

097 Yellow stain from
Inonotus hispidus

Basswood (*Tilia
americana*)
098 Green stain from
Chlorociboria aeruginascens

Horse chestnut (*Aesculus
hippocastanum*)
099 Pink-and-blue stain
from *Scytalidium cuboideum*

100 Yellow stain and zone
lines from *Inonotus hispidus*

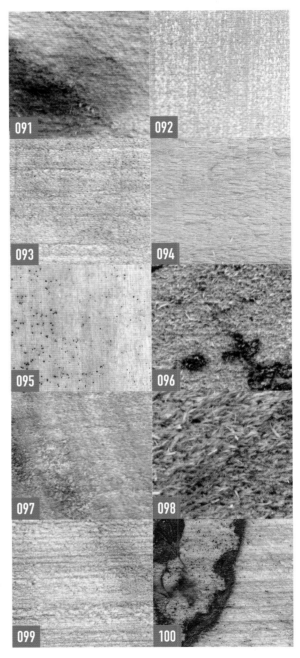

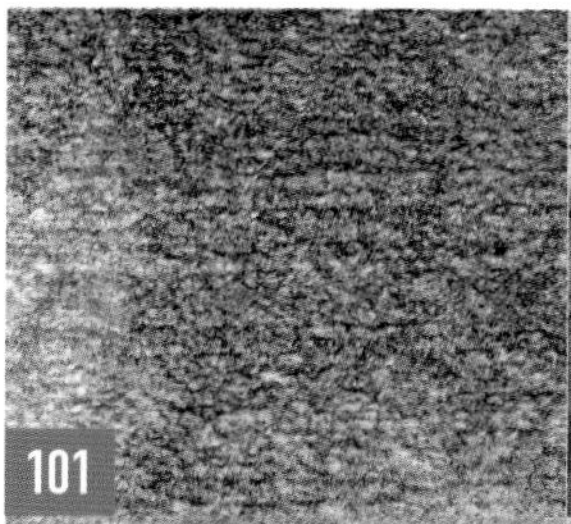

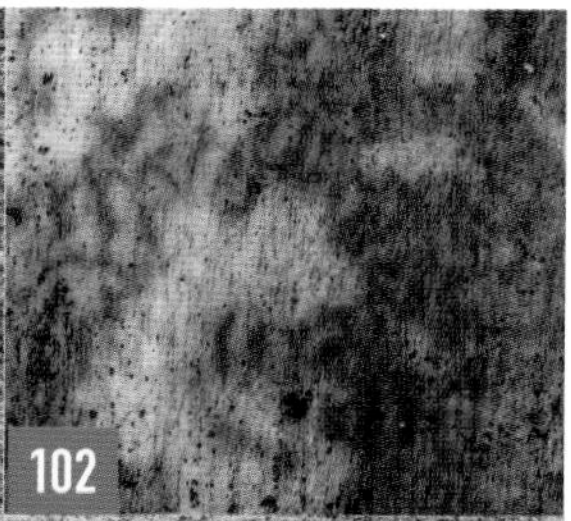

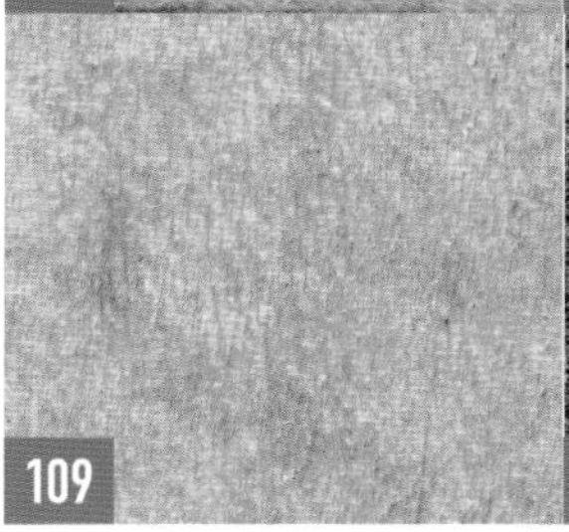

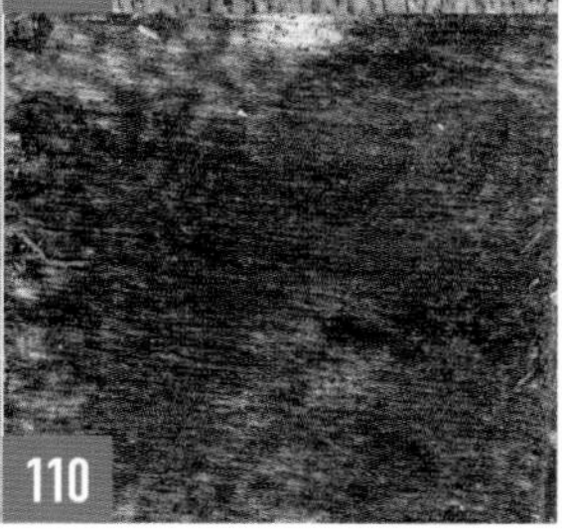

101 Blue stain from *Ophiostoma picea*

102 Zone lines from *Xylaria polymorpha*

Maple, Norway (*Acer platanoides*)
103 Yellow stain and zone lines from *Inonotus hispidus*

104 Blue stain from *Ophiostoma picea*

105 Zone lines from *Xylaria polymorpha*

106 Pink stain from *Scytalidium cuboideum*

Maple, silver (*Acer saccharinum*)
107 Yellow stain and zone lines from *Inonotus hispidus*

108 Blue stain from *Ophiostoma piceae*

109 Pink-and-blue stain from *Scytalidium cuboideum*

110 Zone lines from *Xylaria polymorpha*

Maple, sugar (*Acer saccharum*)
111 Zone lines and blue stain resulting from the interaction of *Xylaria polymorpha* and *Ceratocystis virescens* on copper sulfate–treated wood

112 Blue-and-yellow stain from *Scytalidium lignicola*

113 Pink stain from *Monascus pilosus*

114 Pink stain from *Monascus ruber*

115 Yellow stain from *Scytalidium ganodermophthorum*

116 Zone lines from *Xylaria polymorpha*

117 Pink stain, blue stain, and zone lines resulting from the interaction of *Xylaria polymorpha* and *Scytalidium cuboideum* on copper sulfate–treated wood

118 Zone lines and blue stain resulting from the interaction of *Xylaria polymorpha* and *Ceratocystis virescens* on copper sulfate–treated wood

119 Pink zone lines and pink stain resulting from the interaction of *Scytalidium cuboideum* and *Xylaria polymorpha* on copper sulfate–treated wood

120 Yellow stain from *Scytalidium lignicola*

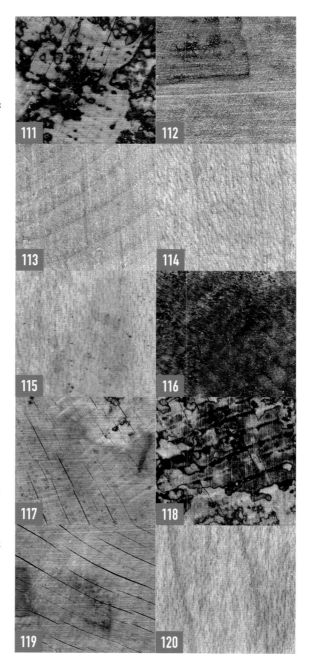

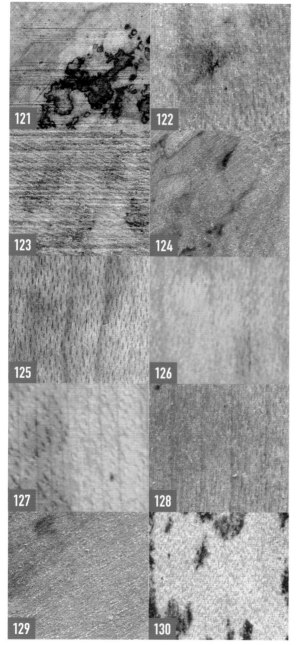

121 Zone lines, blue stain, and pink stain resulting from the interaction of *Xylaria polymorpha* and *Scytalidium cuboideum* on copper sulfate–treated wood

122 Green stain from *Chlorociboria aeruginascens*

123 Blue stain from *Ophiostoma piceae*

124 White rot and zone lines resulting from the combination of *Trametes versicolor* and *Polyporus brumalis*

125 Pink-and-orange stain from *Fusarium reticulatum*

126 Pink-and-blue stain from *Scytalidium cuboideum*

127 Blue-and-pink stain from *Scytalidium cuboideum*

128 Yellow stain from *Scytalidium ganodermophthorum*

129 Blue stain from *Scytalidium lignicola*

130 Zone lines from *Xylaria polymorpha*

131 Yellow stain and zone lines from *Inonotus hispidus*

Tree of heaven (*Ailanthus altissima*)
132 Zone lines from *Xylaria polymorpha*

133 Blue stain from *Ophiostoma piceae*

134 Pink-and-blue stain from *Scytalidium cuboideum*

Chilean woods
Radiata pine (*Pinus radiata*)
135 Red stain by *Eurotium* sp. *Photo by Felipe Galleguillos*

136 Blue stain by *Ophiostoma* sp. *Photo by Felipe Galleguillos*

137 Blue stain by *Phialocephala* sp. *Photo by Felipe Galleguillos*

Peruvian woods
Marupa (*Simarouba amara*)
138 Blue stain by *Cladosporium herbarum*. *Photo by Sarath M. Vega Guiterrez*

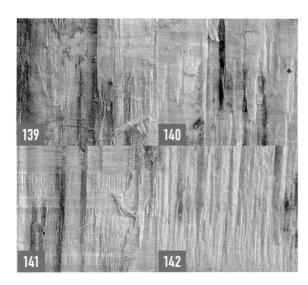

139 Blue stain by *Nigrospora sphaerica. Photo by Sarath M. Vega Guiterrez*

Sapote (*Matisia cordata*)
140 Blue stain by *Cladosporium herbarum. Photo by Sarath M. Vega Guiterrez*

141 Blue stain lines by *Lasiodiplodia theobromae. Photo by Sarath M. Vega Guiterrez*

142 Blue stain by *Nigrospora sphaerica. Photo by Sarath M. Vega Guiterrez*

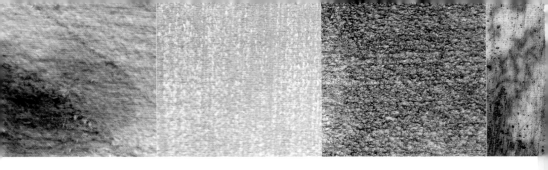

Dr. Seri C. Robinson is a professor of wood anatomy at Oregon State University and an avid woodturner. When not scouring the Amazon for new spalting fungi, Robinson can be found teaching woodturning to eager undergraduates, roller skating, or extolling the beauty of spalted wood. They live with their partner and child in the Pacific Northwest and maintain a fridge filled with fungal cultures and, occasionally, food.
patreon.com/spalting
www.northernspalting.com